R. W. Felkin

On the Geographical Distribution of Tropical Diseases in Africa

R. W. Felkin

On the Geographical Distribution of Tropical Diseases in Africa

ISBN/EAN: 9783744755887

Printed in Europe, USA, Canada, Australia, Japan

Cover: Foto ©berggeist007 / pixelio.de

More available books at **www.hansebooks.com**

ON THE

GEOGRAPHICAL DISTRIBUTION

OF

TROPICAL DISEASES IN AFRICA

ON THE
GEOGRAPHICAL DISTRIBUTION
OF
TROPICAL DISEASES IN AFRICA

WITH AN APPENDIX
ON A NEW METHOD OF ILLUSTRATING
THE GEOGRAPHICAL DISTRIBUTION OF DISEASE

BY

R. W. FELKIN, M.D., F.R.S.E., F.R.G.S.

LECTURER ON TROPICAL DISEASES AND CLIMATOLOGY,
SCHOOL O. MEDICINE, EDINBURGH

WITH TABLE AND MAP

EDINBURGH
WILLIAM F. CLAY, 18 TEVIOT PLACE
1895

EDINBURGH: PRINTED FOR WILLIAM F. CLAY
18 TEVIOT PLACE

LONDON: SIMPKIN, MARSHALL, HAMILTON, KENT, & CO., LIMITED

[*All Rights reserved*]

PREFACE.

As mentioned in the introductory paragraph, these pages, treating of the climatology of Africa and the distribution of disease in that continent, were written at the request of the Committee of the African Ethnological Congress, which assembled at Chicago in 1893.

An abstract of the paper was read before the Congress, and it was published in full in the *Proceedings of the Royal Physical Society of Edinburgh*, vol. xii., part 2.

At the request of many interested in the subject, I venture to publish this very brief contribution to the literature upon tropical disease in Africa.

The text stands as it was written in the early part of 1893, but the map and chart have been corrected up to date.

The Appendix — a paper read before the Congress of Hygiene and Demography at Budapest, in 1894—treats of a new method of illustrating graphically the distribution of disease in, and the climatology of, any area.

I beg to acknowledge my indebtedness to numerous authors from whom I have gained information, in regard especially to those parts of Africa with which I have no personal acquaintance.

R. W. F.

8 ALVA STREET,
EDINBURGH, *January* 1895.

INDEX.

	PAGE		PAGE
Abyssinia,	22, 23	Egyptian Delta,	21
Acclimatisation,	4	Elephantiasis arabum,	62
Africa—		Enteric fever,	67
Altitude of,	6	Equatorial Africa,	37-42
Characteristics of natives,	9, 10		
Emigration to,	11	Gold Coast,	26
Ethnological distribution of population of,	9	Guinea-worm,	57-59
Fauna,	8	Hæmoglobinuria,	75
Geology of,	5		
Lakes of,	6	Lagos,	26
Mortality in tropical districts of,	14	Leprosy,	63
Population of,	9	Madagascar,	45
Rainfall,	8	Malaria,	70-75
Religion,	10	Mashonaland,	43
Temperature,	7	Matabeleland,	43
Vertical zones of climate in,	12	Mauritius,	46
Zones of vegetation,	8	Morocco,	18
African climate—		Mortality in West Coast,	14
Constitutions most suited for,	49, 50	Natal,	44
Effects on emigrants,	15	Native treatment of disease,	47-49
Effects on women,	16	Niger district,	27
African date-mark,	54	Orange Free State,	43
Ainhum,	51	Oriental sore,	54
Algeria,	19		
Altitude, effects on disease,	12, 13	Phthisis,	23, 40, 44
Benguela,	27	Red Sea Coast,	21
Beri-beri,	51	Remittent fever, see Malaria.	
Bilharzia hæmaturia,	52, 53		
Black-water fever,	75	Sahara,	37
		Senegambia,	26
Cape Colony,	44, 45	Seychelles,	46
Characteristics of natives,	9, 10	Sierra Leone,	26
Climate, vertical zones of,	12	Snake bites,	60
Congo,	27	Soudan,	37
		South Africa,	42
Delta, of Nile,	21		
Dengue,	54-57	Transvaal,	24
Disease, influence of altitude on,	12, 13	Tunis and Tripoli,	20
Disease, native treatment of,	47-49	Typhoid fever,	67
Diarrhœa,	68-70	West Coast of Africa,	27-36
Dysentery,	68-70		
		Yaws,	61
East Coast of Africa,	23	Yellow fever,	64-67
Egypt,	21, 22		
Egyptian chlorosis,	59	Zanzibar, vide East Coast.	

ON THE

GEOGRAPHICAL DISTRIBUTION OF TROPICAL DISEASES IN AFRICA.

The origin of this small work was a request from the committee of the African Ethnological Congress, at the World's Fair in Chicago, that I should lecture for them on "Disease and Medicine in Africa." In preparing that lecture, I was desirous of illustrating the climatology of Africa and the distribution of disease on a map, and the one which accompanies this little work was the result of my endeavours. Few maps have been published illustrating the geographical distribution of disease, and, in the one I have constructed, I have attempted a somewhat new departure. I have divided Africa into eight regions, each having approximately the same climatology, and in each of these zones I have introduced symbols to demonstrate the diseases present. I have also endeavoured, by repeating the symbols, to show the prevalence or the importance of the disease. Where one symbol occurs, the disease it represents is present in the area; if two symbols are marked on the map, the disease is very prevalent; and where three symbols occur, the disease is exceedingly rife, or the mortality from it excessive. A mere glance at the map will show the amount of disease in any area, and the comparative salubrity or unhealthiness of a district is at once seen. This, I trust, will be useful, especially as the rush to Africa still continues, and extensive schemes of colonisation are in the air.

When the map I have prepared is compared with one showing altitude, the reason for the presence or absence of many diseases is at once understood.

After describing the climatology of the various regions

into which I have divided Africa, and enumerating the diseases met with in each area, I have added a short description of the tropical diseases met with, adding some remarks upon their probable origin, their prevention, and treatment. This, I think, may be of some use to travellers and others visiting the continent.

Notwithstanding the progress which has been made in medicine, in climatology, and in economics, even now we know little or nothing as to the effects of climate upon the human organism. The subjects of acclimatisation, acclimation, and the survival of the fittest, are, it is true, touched upon by many authors, and it seems to be the idea of many that these problems are more or less settled. This, however, is not the case.

As a rule, authors writing on Anglo-Saxon extension proceed on the *à priori* assumption that the Anglo-Saxon race can live and thrive more or less all over the world; and again, they assume that wheresoever this race may dwell they will remain, morally and physically, as they previously existed in their original habitat. True, some wiser writers do recognise the patent fact that Europeans and individuals from the Northern States of America cannot compete with the coloured races in the Tropics. They allow that the supremacy of the whites in the Tropics is artificial, wherever it may obtain. There the white races only keep *en evidence* practically by constantly reinforcing their ranks through importation. Even if we examine into the condition which obtains in Barbadoes, where the British have their oldest and, possibly, their most healthy tropical possession, we find that a cycle of constant fresh importation has been going on to keep up the stock. It is true that "mean whites" may indeed be found there, but they are the degenerate representatives of the race. In the Southern States of America, good authorities tell us that the Negroes are, owing to their better adaptation to the climate, gradually, but surely, replacing the whites. Looking for a moment to New Zealand, which is, of course, not a tropical country, we find that, notwithstanding the fact that the population is constantly receiving fresh British blood, it differs in a marked degree from the parent stem, owing to the apparently inscrutable action of climate upon human beings. It is evident, even to superficial observers, that the New Zealanders are mentally different from the original stock, and there is, doubtless, also some amount of physical alteration which, were sufficient attention paid to the subject, would be found to be important.

It would be interesting, did opportunity allow, to indicate, with some attempt at detail, the causes which lead to these marked changes, which are noticeable in emigrants and their descendants, but one remark must suffice; the hard winters in the north render the people industrious, provident, and capable of great endurance, or in such a country as New Zealand life is less gloomy and anxious, and the people are more lively—brighter—in fact more mobile. When, however, the climate, like that of Africa, is enervating, emigrants from the north become lazy, indolent, to some extent emasculated. National character is, on the whole, undoubtedly changed. It will be seen in the sequel what conditions to a great extent produce this change in character, and in the Tropics a change in physical stamina, likewise. It will not be possible, however, for me to apply the facts I allude to in this connection. My scope is a more limited one, but I have thought it well to call attention to the subject because of its great importance.

Looked at in the light of modern geology, Africa is of great antiquity. The origin of the continent belongs, doubtless, to Archæan, Palæozoic, and early Mesozoic eras. In round numbers the continent has an area of 9,858,000 geographical square miles, its length is 4330 geographical miles, and its breadth about 5000 miles. It is situated between latitude 37° 20′ 40″ N. and latitude 34° 49′ 15″ S. The coast-line is remarkable; it is only some 17,700 miles in length. This is out of proportion to its vast area, and is due to the absence of bays, inlets, or estuaries. The configuration of the continent of Africa is peculiar. It may be described as possessing a coastal girdle, having a varying breadth of from 100 to 300 miles in width. This girdle bounds an enormous plateau, which slopes from the east towards the west, and in the main axis of which, running from S.W. to N.E., there is a range of mountains, broken, it is true, at intervals.

The main elevation of the continent is less than that of Europe or Asia; an oblique line drawn from Loanda to Suakim passes through a table-land varying in height from 3000 to 4000 feet. It is remarkable that the three great rivers—the Nile, the Congo, and the Zambesi—have their headwaters almost together, so low is the watershed; and the whole river system of Africa renders ready drainage impossible, and thus gives to it, on the whole, a water-logged character, especially in vast areas in the centre of the continent. It is noteworthy that there are six drainage areas in Africa, three inland — the Sahara Desert, the Kalahari Desert, and Eastern Abyssinia — and then the Atlantic Ocean, the Mediterranean Sea, and the Indian Ocean,— the drainage area towards the Atlantic Ocean, including as it does the river systems of the Niger and the Congo, being the greatest. With the exception of the Nile, whose discharge is only about one thirty-seventh of the rainfall of the river's basin, the African rivers have an abnormal volume of discharge, due to the situation of their mouths in the equatorial zone, and the excessive rainfall in their catchment basins.

The lakes of the African continent may be divided into three classes—those situated in the continental axis, namely, the Victoria, Albert and Albert Edward Nyanzas, Tanganyika, and Lake Nyassa, possessing fresh water, and having an oceanic outlet; secondly, Lakes Chad and Ngami, which lie in continental depressions, and vary greatly in size during the various seasons, whose water is more or less saline; and lastly, the lakes situated along the course of the great rivers, and which are simply enormous expansions of these, caused by the heavy rainfall in the Tropics. In Central Africa the highest mountains are Kenia and Kilimanjaro in the east, and Gambaragara and Ruwensori in the centre, having an average altitude of some 19,000 feet.

Our knowledge of the climatology of Africa, except in northern and southern districts, is very incomplete, and this is unfortunate, because an exact knowledge of the climatology of Central Africa would be of the utmost interest and importance to us in our present inquiry. It will, however, probably be of advantage, before dealing with the climatology of distinct areas, which must, of necessity, be treated separately, to give a summary of the general phenomena of the climate of the whole continent, so far as it is known.

The mean annual temperature of the continent of Africa is high. There is a strip of country upon the east coast from lat. 23° N. to lat. 23° S., where the mean annual temperature is over 80° F., and the same high temperature obtains in two other regions. One of these is towards the west of the Red Sea, bounded towards the north by Dongola, on the south by Lado on the Nile, on the west by about long. 25°, and on the east by Abyssinia and long. 36°. The other area is situated towards the west, bounded on the south by the coast-line between Sierra Leone and Lagos, on the west by long. 17° W., on the east by long. 18° to 20° E., on the north by a pyramidal line lowest on the west and east at lat. 16° N., and having its apex at lat. 28° N., long 0°. Apart from this, with the exception of Morocco, Algeria, Tripoli, and the Nile delta, an oblong area commencing in Abyssinia, and running down to the equator, Cape Colony, Matabeleland, Damaraland, and Angola, where the mean annual temperature is between 60° and 70° F., the remaining part of the continent has a mean annual temperature of from 75° to 80° F. From this it will be seen that, although along the northern, western, and southern coast-line the temperature is to some extent lowered by the proximity of the ocean, yet on the east the sea exerts no modifying effect, but practically increases the temperature 5° to 10° F.

With regard to the mean annual range of temperature in Africa, considerable variations exist. There is a district which may be roughly said to embrace nearly the whole of the Congo Free State, where the range is less than 5° annually, and in the Sahara we arrive at the other extreme, the range being there between 40° to 50°. Practically, from lat. 18° N. to 20° S., the annual range is from 10° to 20°. With regard to relative humidity, it is over 70 per cent. over a third of the continent. The upper boundary of this area may be described as a line drawn from Cape Verd to Lado on the White Nile on the north, bounded by 32° E. long. in the east and lat. 15° S. in the south, and including, also, the coast belt from Cape Town to lat. 2° N. on the east coast. Apart from this, with the exception of Abyssinia, where the humidity is also over 70 per cent., it is considerably less.

The rainfall in Africa likewise varies greatly, it being under 5 inches in the Sahara and in the Kalahari Desert. It is over 100 inches at Sierra Leone and between Lagos and Gaboon. Over the remaining part of the continent it varies from 10 to 100 inches. Speaking generally, the annual rainfall may be estimated at from 50 to 60 inches. The rainfall at the equator is pretty evenly distributed throughout the whole year, but to the north of the equator, at about 15° lat., there are two well-marked rainy seasons.

With reference to winds, the trade winds are the most important. North of the equator the prevalent trade winds are from the north-east, south of it from the south-east, with an intervening belt of calms at the equator, and it is to these winds that the unequal distribution of the rainfall is to be attributed.

The heat, rainfall, and winds naturally affect the fertility of the country, and the whole area of Africa may be divided up as follows:—36·4 per cent. is occupied by deserts, 14·6 by steppes, 5·3 by scrub, 21·3 by savannahs, 21·8 by forests and cultivated land, and 6 per cent. by the large lakes. Therefore half the continent is occupied by deserts and steppes.

The typical zones of vegetation are—(1) the Mediterranean zone, having a vegetation similar to that of southern Europe, if somewhat more tropical; (2) the Sahara Desert zone, of which much might be cultivated, numerous oases certainly existing there; (3) the zone of tropical vegetation, more or less coincident with the areas of the highest mean annual temperature and heaviest rainfall; and lastly, the south-central and South African savannah zone. In general terms, the vegetation is richer as we proceed from the south to the north, and also from the west to the east.

It is unnecessary to refer in detail to the fauna of Africa. The mammalian fauna is exceptionally varied, the bird fauna meagre, the reptile fauna largely developed. Insect life is very abundant, and the tetze fly infests large parts of the country which are uncleared.

The population of Africa may be estimated at about 20 to a square mile. Boehm and Wagner estimate the population at 205 millions, Mr Ravenstein at 127 millions, with a rate of increase of 10 per cent. per decade.

The people of Northern Africa were probably in prehistoric times of the same ethnical stock as that of the people inhabiting Southern Europe. The Arabs of the Soudan are probably all descendants from the proto-Semitic stock. For the rest we have Negroes of various kinds, but all distinguished by projecting jaws, flat features, broad noses, woolly hair, shining skin, and pouting lips, and it is probable that they are all of one origin. We find also in Africa the Bushman and Tikki-Tikki or Akka dwarfs, all of whom I believe to be among the oldest primitive inhabitants of the continent.

In classifying the ethnographic distribution of a population, it is best to be guided by linguistic facts; but, as according to Cust there are no less than 438 languages and 153 dialects spoken in Africa, such a classification is for present purposes impossible, and therefore I adopt Miller's classification of distinct groups, as follows:—(1) The Semitic family, along the north coast of Africa and of Abyssinia; (2) the Hamitic family, who dwell mainly in the Sahara, Morocco, Algeria, Egypt, and in the Galla and Somali districts; (3) the Fulah and Nuba groups, who live in the western central and eastern Soudan; (4) the Negro groups, in the western and central Soudan, in Upper Guinea and the Upper Nile region; (5) the Bantu family, everywhere south of 4° N. lat., except in the Hottentot domain; (6) the Hottentot group, in the extreme south-western corner of Africa, from the Tropic of Capricorn to the Cape; (7) the Tikki-Tikkis and Akkas, living in scattered groups to the north of the equator.

A few sentences with regard to the general characteristics of these people are necessary, to show their possible predisposition to certain diseases, their immunity from others. The inhabitants of Africa have, on the whole, a well-developed muscular system. Mentally they are like children, easily amused and easily roused to passion. They possess innate capabilities of high education, but it is a mistake to suppose that the average Negro child can be educated up to the European standard. Till the age of fourteen he will probably distance a European child in almost all brain work, but after this age the light-skinned Caucasian shoots ahead of the dusky child of the Tropics. As a rule, at least three generations are necessary to develop the Negro to our mental standard. Owing to the climatological factors to which I have referred, the natives possess a lethargic constitution. Nature being so bountiful, they have no incentive to strenuous manual labour. But there is a marked difference between the inhabitants of the mountainous regions of the Tropics and those inhabiting the low-lying plains. For instance, the inhabitants of the northern and higher parts of Uzinza are far more strongly built and energetic than the Wazinza who inhabit the southern and lower parts of the country. Such, at any rate, was the opinion of Speke, and I have personally noticed the same characteristics north of the equator.

Again, the religion and customs of the inhabitants necessarily influence their susceptibility to disease, or their power of recovery after having been attacked by disease. There is a marked difference between the stolid, fatalistic, Mohammedan and the superstitious pagan. A difference also obtains in many districts owing to diet, partly influenced it is true by custom, but also by religion, and the vegetarians and mixed feeders are liable to definite varieties of disease.

Although it is a mistaken notion that insanity and nervous diseases do not obtain among the natives, yet they are undoubtedly less frequent in Africa than they are in temperate latitudes; and it is certain that civilisation, or, at any rate, contact with the whites, has a more or less deteriorating effect upon the native population.

It is now necessary to refer briefly to the subject of acclimatisation, and to endeavour to answer the question which is frequently put—Can Europeans become acclimatised in Tropical Africa?

My strong opinion is that it can only be possible if migration occurs step by step, and in estimating the possibilities of acclimatisation we must count by generations rather than by years. I believe, however, that our increased knowledge of climatology and hygiene renders the problem of acclimatisation more easy of solution than it was, and, given picked individuals and the careful selection of tropical areas in which to colonise, I see no reason why, with precautions, inhabitants of the temperate zone should not colonise even in Central Africa. In the selection of emigrants, great care should undoubtedly be taken, and all persons with a tendency to gout or rheumatism, diabetes or albuminuria, those with a nervous or alcoholic family history, or those suffering from either acquired or hereditary syphilis, should certainly be restrained from emigrating to Tropical Africa.

As I before indicated, environment definitely influences not only individuals but nations, and for this reason a marked difference obtains between the adaptability of residents in the northern and southern parts of a temperate zone for acclimatisation purposes. Individuals and nations are not only influenced by the climatology of their original residence, but also by their habits and customs and their psychical peculiarities. It has been found, as was pointed out long ago by Mr Ravenstein, that the peoples of Southern Europe, such as the Italians and southern Frenchmen, can withstand the climate of sub-tropical Africa better than can northern Europeans. "A steady stream of migration is in fact setting in, in that direction. Germans and Belgians are pouring into France; Frenchmen are going to Algeria; the Arabs from the shores of the Mediterranean have found their way into the Soudan; whilst the Soudanese are pushing forward into Bantu Africa. A similar movement is going on in South Africa. The descendants of those Dutchmen who, a couple of hundred years ago, first settled at the Cape have made their way to the Transvaal; and European migration, favoured by geographical features, is being pushed even within the Tropics towards the Zambesi."

This agrees with my own opinion, that wholesale immediate acclimatisation for Europeans in Tropical Africa is entirely out of the question.

But it may be said, What about the high African tableland? What about the mountainous regions? As it is customary to speak of three climatological zones—the hot, the temperate, and the cold—between the equator and the poles, so in like manner we can say that in the Tropics there are three vertical zones of climate. (1) A zone extending up to a height of 3000 feet, having a mean annual temperature of from 72° F. to 83° F.—the hot zone; (2) a zone from 3000 to 12,000 feet, with a mean annual temperature of from 41° to 72° F.—the temperate zone; (3) a zone from 12,000 to 16,000 feet or above, with a mean annual temperature of from 30° to 41° F.—the cold zone. Each zone has peculiarities of its own with regard to the presence or absence of disease, and the higher the altitude of the region within certain limits, the more nearly it approaches the climate of Europe. As we proceed, however, north or south of the equator, the boundaries of those zones are found at gradually decreasing elevations, and therefore heat, atmospheric pressure, etc., vary at different latitudes.

The influence of these zones upon disease is, broadly speaking, as follows:—In the upper or cold zone there is a tendency to plethora; the disorders met with are of an inflammatory character, and the diseases of the respiratory and circulatory organs are far from uncommon. Malaria, yellow fever, cholera, phthisis, dysentery, and hepatitis are almost entirely absent. The absence of phthisis is probably due to the rarification of the atmosphere, the absence of yellow fever to the lower temperature, as well as on account of the distance from the sea.

Areas situated at these high altitudes are incomparably superior to the low-lying districts. As a rule they are entirely free from tropical endemic diseases, which, should they perchance be introduced, rarely spread. A marked difference is soon noticed in the appearance of individuals who, after a residence in low-lying tropical countries, go to the hills or elevated tablelands. Their vigour improves rapidly, they almost regain the healthy appearance they had in Europe, their digestion and the composition of their blood is improved, and a proper amount of exercise can be taken.

"Hill diarrhœa" and rheumatism may sometimes affect them in these higher altitudes. The former may be due to either a faulty water-supply or a too rapid removal from the plains; the latter—as in India—to damp dwellings, or to the more marked vicissitudes of climate.

The middle zone is perhaps best divided into two, which we may compare with the temperate and sub-temperate zones, the cooler (higher) having a mean annual temperature of from 41° F. to 55° F., the warmer (lower) a mean annual temperature of from 55° to 73° F. In both these areas the seasons exert an influence, and therefore the presence and prevalence of disease fluctuate, the frequency of diseases of the respiratory and digestive organs, for example, rising and falling correspondingly; but throughout this region, as a whole, diseases specially frequent in the higher and lower zones are less virulent in character.

The lowest or hot zone is the typical tropical disease zone. Here we meet with anæmia, malaria, diseases of the gastro-intestinal tract, hepatitis, dysentery, diarrhœa, beri-beri, dengue, yellow fever, etc. Various diseases, however, are not met with, such as typhus, plague, goitre and cretinism, and, for the most part, diseases of the kidneys. It is obvious that inhabitants of the temperate zone dwelling in this lowest or hot zone must be injuriously affected by the climate, and probably the greatest injurious effect they sustain is due to the heat and equable temperature, the tonic effects of the cold season being sorely missed, in consequence of which there is a gradual sinking of the vital energy. Moisture is the next injurious element, for dry heat is much less injurious to such emigrants than a hot moist atmosphere. The character of the soil exerts a not inconsiderable influence, low-lying clayey soil, soil rich in mould, and alluvial soil, acting injuriously; dry or sandy soil being less injurious. If the soil be marshy, temperature is lowered, and it is found that by draining the soil the temperature will be raised 2 or 3 degrees. The drier the soil the greater the heat during the day, and the greater the cold by night, owing to the rapidity with which the soil cools by radiation. Vegetation also influences the salubrity of these regions, for where it is abundant, the temperature will be more equal, the vegetation preventing the sun from heating the ground, and also preventing radiation during the night. Lakes also exert an influence, by preventing much variation between the temperature of day and night.

A good deal is written concerning the high rate of mortality in tropical regions, but in comparing the salubrity of areas in the temperate zone and in the tropics, it is often left out of mind that much of the salubrity of Europe is due to artificial causes—to the prevention of disease, to sanitary measures, to the care of the sick and infirm, and to the superior medical knowledge of our law and civilisation. This fact can be at once appreciated when we consider the state of health and rate of mortality in different European states; and when we find such marked contrasts as are presented in Great Britain and Russia for instance, we observe that locality alone does not cause the great difference, and that it is not climate alone which influences disease and mortality.

The general salubrity of any place may be ascertained from the death-rate; in fact it is the only criterion we have. A low death-rate—say 20 per thousand—indicates that the climate and sanitary surroundings must be good. If the death-rate be, say 60 per thousand, either climate or sanitation must be at fault. In speaking of a "bad" climate in Africa, it must not be forgotten that what is meant is that the climate is bad for emigrants from the temperate zone, not necessarily that it is bad for the natives. Take, for instance, the west coast of Africa. The climate is not on the whole fatal to natives, and their death-rate is not immoderately high, although the district goes by the name of "the white man's grave." There are, it is true, exceptional areas where even the mortality amongst natives is great, but so too in the temperate zone there are areas where, owing to faulty sanitation, overcrowding, and the like, a state of matters inimical to health, and inducing a high rate of mortality, is to be found. A few statistics, taken haphazard, will show what I mean. The death-rate in Edinburgh in 1887 was 19·8 per thousand; in Manchester in the same year, 28·7; in New York, 1878-80, 26·2; in Chicago, during the same period, 27·2; in St Louis, 19·3; in Calcutta, 31·1.

Very much may be done to render even the worst climate in Tropical Africa more salubrious, and the sanitary precautions, to which attention will be called in the sequel, will do a great deal to raise the health of the community. For instance, attention to the water-supply at Sierra Leone has had a very marked effect in lowering the death-rate; but when all is done, a permanent residence for inhabitants from the temperate zone is at present out of the question in the low-lying regions of Tropical Africa.

We must now briefly consider the effects which the African climate exerts upon emigrants from the temperate zone, and which are chiefly the results of heat and moisture. For a short time after the arrival of such an emigrant in Tropical Africa his health remains good, in fact the heat seems to have a stimulating effect upon him. He does not suffer from the heat of the sun, and he forms a striking contrast to those emigrants who have resided in the same district for a considerable time. His work, be it mental or bodily, is performed with comparative ease. Soon, however, he begins to experience a marked change in the physical functions of his body. Its normal temperature is about 1° F. higher than it was in the temperate zone, his respirations are considerably below the normal number, his heart's action is increased in frequency, his digestive powers become weakened, his skin secretes far more freely and becomes somewhat swollen, his urine is lessened in quantity, and his nervous system is, to a greater or less extent, enfeebled. There is no doubt that very soon his mental and bodily powers become weakened, this being due to modified nutrition. He suffers from anæmia and from a slightly congested liver. The anæmia, if only slight, may be considered as prophylactic in character, and the congestion of the liver is only what might be expected on account of the increased work that that organ is compelled to perform. The resistance which an individual from the temperate zone can offer to these conditions depends on the height of the mean annual temperature, and also to some extent on its diurnal variation. If the temperature is high during the whole year, he is less able to withstand them than he would be were there marked differences in the temperature of the various seasons, or between that of night and day. This power of resistance may also be attained by a change of air, at stated intervals, to a place, either having a high altitude, or where marked diurnal variation in temperature occurs. The difficulty experienced in Africa at the present time is, that comparatively few high altitude stations are known, and some considerable number of years must elapse before arrangements can be made to provide access to those areas where the emigrant, enervated by the heat and moisture, may seek the restoration provided by a residence at a high-level station.

A word in passing as to the influence of the African climate upon women. Practically, although women do not suffer so much as men from malarial fever, this is probably due to their not being so much exposed to its influence. Taking, however, all things into consideration, I have come to the conclusion that neither sex is more capable than the other, of withstanding the climate. With regard to children, it may be said that the same condition obtains in Africa as in India. Apart from accidental diseases, children will do well for the first three or four years of life, but under existing circumstances they must then be sent to a temperate climate if they are to survive. Were they to remain in Tropical Africa, they would certainly degenerate mentally, morally, and physically.

I have now come to the end of my extremely rapid survey of the general facts which obtain in Africa with relation to the subject of my paper, and I now proceed to describe, solely from the point of view of health and disease, the various regions into which Africa may be divided, and to indicate the special tropical diseases which obtain in them. When, then, the area of the distribution of the diseases of Africa has been geographically defined, I shall proceed to briefly examine each tropical disease, the conditions under which it arises, and its prevention and treatment, *seriatim*.

For my present purpose I can divide the continent of Africa into eight distinct divisions, namely,—

   I. Northern Africa, including Morocco, Algeria, Tunis, and Tripoli.
  II. North-Eastern Africa, including Egypt and Abyssinia.
 III. Eastern Africa, including the Islands of Zanzibar and Pemba.
  IV. West Coast of Africa.
   V. North Central Africa, comprising the Sahara and the Soudan.
  VI. South Central Africa, extending to about 18° S. latitude.
 VII. South Africa.
VIII. The Islands of Madagascar, Mauritius, and The Seychelles.

I. NORTHERN AFRICA.

This part of Africa has its climate modified by its proximity to the Mediterranean and Atlantic, and also by the ranges of mountains running parallel to, though at some distance from, the coast. It naturally falls into various political divisions.

1. *Morocco* is a dry and healthy country. It has a moist and fairly equable climate. There are two seasons—September to May being the winter, in which the rainfall is about 30 inches, and the temperature varies from 50° F. to 65° F. The summer, almost rainless, has a temperature of 65° to 80° F. Throughout the whole year the daily variation of temperature is remarkably slight.

The diseases specially met with in Morocco are dysentery and diarrhœa, which are very prevalent along the coast; and leprosy, syphilis, and ophthalmia, which are very common throughout the whole country. Chronic rheumatism is prevalent, scrofula not uncommon, but respiratory diseases and the eruptive fevers, with the exception of small-pox, are rarely seen. Malaria, manifested by intermittent fever, only obtains in a slight degree. Epidemics of cholera have frequently visited this country, as also all the countries included in Northern Africa.

2. *Algeria.*—This country must be divided into two parts—the sea-coast district, called the Tell, and the elevated region beyond, extending to the summits of the mountains, which have an altitude of from 3000 to 8000 feet. The mean annual temperature is about 65° F., the mean diurnal variation 42° F., the nocturnal variation 36° F. The average relative humidity is about 45 per cent. The rainfall varies in different districts, but the average at the coast is about 900 mm. The rain commences late in October, and ends about the end of March, autumn and winter occurring within this period. Spring begins in the middle of March and ends in June; July, August, and September being the summer months. There is a greater amount of disease in Algeria than in either Morocco, Tunis, or Tripoli. The death-rate from typhoid fever is slightly over 11 per 1000, this

disease being extremely common, especially in the towns.
Diphtheria and croup are also widely prevalent, being 15 per
1000; diarrhœa and gastro-intestinal disorders rather over 29
per 1000. Dysentery also is very prevalent, especially in the
district of Oran, where it is said to be due to the special
character of the water. Small-pox and measles are to
be found in Algeria, and both acute and chronic bronchitis
and pneumonia are very frequently met with. This is contrary to what one might have expected. With regard to
phthisis, the death-rate is high, but this is only in the towns,
and is caused by the number of phthisical patients sent to
Algeria in the hope of cure. It is a great mistake to
send patients, in whom phthisis has become well-developed,
to Algeria, for, although the climate is exceedingly good
for those in whom the disease is either threatening or is
in a very early stage, it is undoubtedly prejudicial to those
in whom it has taken firm hold. Leprosy and Oriental Boil
are extremely common, and so is hepatitis, both the acute
and chronic varieties, and, especially in Oran, tropical
abscess of the liver. Syphilis and acute rheumatism do
not occur nearly so frequently as in Morocco. Goitre is
prevalent in the mountainous regions. Malaria is very
prevalent in Algeria, although it is not so frequently met
with as formerly, for sanitary science is beginning to make
itself felt, and so diminishes its frequency. It is not
so often seen in the province of Algiers itself, as it
is in those of Oran and Constantin, in which districts it is
about twice as prevalent. The amount of fever is unequally
distributed throughout the year; the maximum amount
occurs in the autumn, the minimum between December
and May. July and October appear to be the two months
when fever is most prevalent. Undoubtedly the heavy rainfall from November to February diminishes the fever. On
the whole the fevers are most prevalent in marshy areas with
a sub-soil of clay, but, as Colin points out, there are localities
infested by malarious fevers which are far from being marshy.
But this forms one of the peculiarities with regard to malaria,
and, as I shall point out elsewhere, is, I believe, due to the
height of the ground-water. Probably the most malarious

districts are Mitidja, Bona, the Plain of Eghris, Gigelle, the plains of Zig, Habra, Zeybouse, and of Shotts, also the district through which the Macta Canal passes, and the borders of the Fezara Lake.

3. *Tunis and Tripoli.*—These regions are for the most part deserts, having a mean annual temperature of about 70° F., and a mean annual variation of about 30° F. Diphtheria, dysentery, and diarrhœa are very prevalent, the two latter diseases being especially met with between July and October. Syphilis and acute and chronic rheumatism are also very common, but diseases of the lungs and liver exceedingly rare. Almost the whole of the district is malarious, with the exception perhaps of the area in which is situated the Lake of Bizerta, and also Porto Farina.

Summarising the information we possess with regard to Northern Africa, we find that malaria, dysentery, diarrhœa, leprosy, syphilis, and rheumatism are exceedingly frequent; but it seems only probable that, were extensive sanitary measures adopted in the various countries referred to, and especially drainage and a good water-supply provided, the district would be far from inimical to emigrants from the south of Europe, and the climate is decidedly favourable to persons suffering from incipient phthisis. Such individuals should, however, spend at least two winters and the intervening summer in North Africa, if they would gain any decided benefit from a sojourn there. It is not altogether easy to estimate the progress made by France in the colonisation of Northern Africa, but on the whole I believe that gradual acclimatisation is taking place, and that in time the country may be completely and successfully colonised by that nation; and doubtless Italians and Spanish would fare equally well.

II. NORTH-EASTERN AFRICA.

In this division of my subject there are various climatological areas to be considered—(1) The Delta; (2) the Valley of the Nile; (3) the Eastern Desert, including the coast of the Red Sea; and finally Abyssinia. Part of the country politically known as Egypt will be considered under the head of the Soudan.

1. *The Delta.*—This area is humid, but the annual rainfall is only about 7 inches. Northerly winds prevail. The mean annual temperature is about 69° F., and the diurnal variation slight.

2. *The Valley of the Nile.*—The climate in this district is very different. The country may be practically said to be rainless. At Cairo there are only four or five showers a year, and in Upper Egypt but one or two. The mean annual temperature at Cairo is about 72° F., but the variation between night and day is very great. At Cairo the temperature may be as high as 110° F., but in the winter it may fall to below 32° F. The marked feature in the Nile Valley is the inundation of the country by the Nile, which begins to rise about the middle of June, reaches its height at the beginning of October, when it commences to fall, and it must be remembered that in Upper Egypt the river forms the sole source of the water-supply to the country. There is probably no other country in the world where the population is so dependent upon a single river, and unfortunately the Nile is habitually polluted with all kinds of filth, which has a great effect on the health of Egypt, for during the summer months, when the Nile is low, the people practically imbibe a solution of filth. This faulty water-supply undoubtedly causes the tremendous mortality amongst the child population. Out of 1000 children born in Egypt, 496 die before the age of five years (H. R. Green). It would be quite possible, although the cost would be great, to improve the water-supply; there are no insuperable engineering difficulties to be encountered.

3. The *Eastern Desert* and the *Red Sea* coast are drier and hotter than the Nile Valley itself. This notwithstanding, the country is salubrious. The heat from January to April is almost insupportable, especially during the southern khamseen, which is a dry scorching wind.

With regard to the diseases prevalent in these regions, typhoid fever, relapsing fever, and dysentery are very prevalent throughout the whole country, and simple febriculas are frequently met with. Epidemics of cholera have often visited Egypt, but plague, which used to be endemic here, has not reappeared since 1844. Ophthalmia is extensively prevalent all over Egypt. It is said that about a fourth of the whole population of Egypt is affected by the *Anchylostomum duodenale*, which causes the Egyptian chlorosis. Round worms are also very common, and guinea-worms, introduced from the south, are not infrequently seen. There is also a specific affection of the urinary passages caused by the *Bilharzia hæmatobia*, the embryos of which infest the drinking water. Small-pox is epidemic; measles and scarlet fever are frequently seen, and so are hepatitis and tropical abscess of the liver. Phthisis is very rare, except among the Negro population, in which it is frequently met with. Rheumatism, on the other hand, is frequent, especially at Cairo. Scrofula affects the natives, leprosy is endemic, and syphilis very common. On the whole, the country I am describing is remarkably free from malaria, although the disease is present in the autumn. The greatest number of cases occurs at Suez, where the sub-soil water is very high. Suakim and Massowa are also said to be malarious, but, owing to the character of the fevers in these two places, it is doubtful if we are not dealing with dengue, or may be with an endemic form of influenza.

4. *Abyssinia.*—Abyssinia, which is situated between 7° 30' and 15° 30' N. lat., and between 35° and 40° E. long., has a mean altitude of about 7000 feet. It is an extensive tableland, from which mountains rise to a height of nearly 15,000 feet. The highlands are mostly covered with pasture or cultivation, forests being rare. The climate of the plains varies with altitude, a very large district being healthy and temperate, suited even to European colonisation. The mean annual temperature of the lowest part of Abyssinia is about 86° F., and the rainy season begins in December and ends usually in March. In the highlands the mean annual temperature varies from 67° to 55° F., according to altitude. The rainy season is from June to September, the rainfall being about 40 inches.

Diarrhœa, dysentery, and rheumatism are very prevalent

in Abyssinia. Small-pox, for which, by the way, inoculation is employed, is exceedingly prevalent, as also is syphilis, and leprosy is fairly common. The people, too, suffer considerably from worms, this being due to their custom of eating raw meat. Probably phthisis rarely occurs in Abyssinia, and diseases of the chest are on the whole infrequently met with. In the rainy season typhoid fever, relapsing fever, and epidemic influenza are common. The natives also suffer from scrofula and leprosy, which is especially prevalent in the mountains, and goitre is said to occur. Cholera has frequently visited the country. Malarial fevers occur in Abyssinia, but not to anything like the extent that they do in other parts of Africa. There are, however, a considerable number of places where malaria is endemic.

III. EASTERN AFRICA.

I include in this division of my subject the country between Cape Guardafui and latitude 18° N., and the coast inland to about longitude 36° E.; also the island of Zanzibar, because it really belongs to the east coast of Africa, and the conditions which obtain there are practically those which are found on the coast.

In the northern part of this area, between 10° N. lat. and the equator, we find the Galla and Somali districts. Little is known of them from a climatological and medical point of view, but in all probability, they have the same characteristics as those mentioned when referring to Abyssinia, and these regions are undoubtedly healthier than any on the east coast farther south, if we except the highlands between Kilimanjaro and Mount Kenia and Lake Victoria Nyanza.

From the equator to Delagoa Bay there are two seasons—the dry season from June till October, the wet one from November till May, the greatest rainfall being from April to June. The amount of rainfall varies in different localities from 1500 to 2500 mm. The mean annual temperature is about 80° F. at Zanzibar. February is the hottest month, with a mean temperature of about 84° F.; July the coolest month, with a mean temperature of 77° F. The humidity is very marked.

The inland districts of the east coast region do not vary much from the conditions which obtain at the coast, though of course altitude makes itself felt in modifying those conditions somewhat. Thus, around Kilimanjaro the mean annual temperature may be said to be slightly higher than at Zanzibar, mainly about 85° F., but there is a somewhat greater range, and the nightly mean is probably about 66° F. Farther south in the Mpwapwa district, according to Pruen, the maximum daily temperature during the hot season varies from 80° to 90° F.; the minimum at nights is 65° F. During the cold season the daily maximum varies from 70° to 80° F., the lowest night temperature being 60° F. At Blantyre, still farther south, and at an altitude of 3000 feet, the mean annual temperature is 64° F. The hottest season is during October and November, when the mean temperature is about 75° F., the coolest season being June and July, with a temperature of about 60° F. The rainfall in these districts is the same as that upon the coast, the smallest precipitation being near Blantyre, with an average amount of 58 inches, and increasing to 200 inches in the north.

The east coast of Africa is undoubtedly very unhealthy, and on the coast, and with few exceptions in the interior, emigrants from the temperate zone cannot thrive. Practically, the Shiré highlands, the slopes of Kilimanjaro and the districts to the north-west of that mountain are the only two areas in which it is possible to think of white races colonising. The diseases which do not occur on the east coast are scarlet fever, phthisis, and goitre, nor is cancer met with. Whooping-cough is occasionally known; hepatitis and tropical abscesses of the liver are less frequently met with than might be expected; measles are very rare, although one or two serious epidemics have occurred in the southern part of the district under consideration. Both the anæsthetic and tubercular forms of leprosy are occasionally met with. Cholera has visited the east coast of Africa on several occasions (Christie). Rheumatism, especially the chronic form, is met with all over this area, and syphilis and ophthalmia are very prevalent. *Elephantiasis arabum* is also seen, but not extensively, except at Zanzibar, where it is very common. Tropical ulcers are very frequently seen; they occur chiefly in the debilitated natives, and cause great havoc in the slave caravans. Diarrhœa prevails extensively. Dysentery is extremely common throughout the whole of the area, especially at Zanzibar. Epidemics of dengue are of frequent occurrence. Although, as has been mentioned, phthisis is not met with in East Africa, yet pleurisy and bronchitis are very common. Typhoid fever is fairly common. Beri-beri also exists in East Africa, and it is probable that the epidemic dropsy which is mentioned by Livingstone was really this disease. Throughout all East Africa malarial fevers are very rife; they are most severe upon the coast, along the river valleys, and in marshy, water-logged areas; in fact they are met with wherever the factors known as those producing malaria exist. To these I shall refer subsequently when dealing with the prevention of the disease. It should be noted, however, that the fevers on the table-land are much less fatal in character than those met with upon the coast, and there are areas over 2500 feet in altitude, where it may be said that the incidence of malaria is of but slight import. It may be well to mention here that the reason why East Africa has such an evil repute, owing to so many deaths having occurred amongst the Europeans who have visited it, is due to the fact that these explorers and missionaries have been compelled to make a long stay on the coast before proceeding inland, and have there become saturated, as it were, with the malarial poison. Were these conditions rectified, and were it possible for the whites to proceed rapidly to the interior, there is no doubt that their general health would be far better, and it is for this reason that for the last ten years I have so strongly advocated the construction of short railways to carry travellers rapidly across the malarious belt on the coast.

IV. WEST COAST OF AFRICA.

The west coast of Africa from Cape Verd to Cape Frio, is the hottest and wettest region in the globe, probably the most unhealthy too. The expression, "the white man's grave," which is given to this part of Africa, is well-deserved. The conditions which are so inimical to the white races in this region are continuous heat, excessive moisture, and sudden changes of weather, but much may be done to render the climate more salubrious by sanitary precautions. On the whole coast, probably the Cameroon mountain system is the only place of sufficient altitude for a really valuable health resort, but the Portuguese portion of the coast is not so unhealthy as that farther north.

A rapid summary of the climatological conditions which obtain in the various regions along the west coast must now follow.

First, with regard to the coast of Senegambia. The mean annual temperature is 75° F.; the mean diurnal range 9° F. The average rainfall is about 18 inches; the average humidity 73. The hottest months are June to November. In the interior of the country the mean annual temperature is about 83° F., the hottest months being April, May, and June. There is considerably more variation in the temperature in this inland region, it having an average for the year of 18° F. At Bathurst, which is situated on St Mary's Island, the chief rains fall from July to September, about 39 inches falling in 80 days. At Sierra Leone the mean annual temperature is 82° F.; the rainfall 100 inches in August and September. On the Gold Coast at Accra the mean annual temperature is 80° F.; the mean range about 7° F. Humidity is very high, averaging about 75 per cent. The rainfall varies considerably, some years being as low as 23 inches, in others nearly 40. The rainy season begins at the end of March and continues to June. The dry season is from November to March.

We may take the climate of Lagos, still farther to the east, as another example of the meteorology of Western Africa. The dry season obtains from the beginning of December to the middle of March; then follow the heavy rains until the middle of July; from then till September there is a moderate amount of rain until the end of November. There is however some rainfall during each month, but only about 1 inch from December to March.

The total rainfall is about 58 inches annually, and the mean annual temperature 80° F. The mean annual variation is very slight, being only about 10°.

The Niger district has almost the same characteristics as the one just mentioned, except that the temperature is somewhat higher, the heat during March to June or July averaging 105° F., the mean annual temperature being about 86° F. The rainfall is about 60 inches; the relative humidity about 85. Proceeding farther south, we come to Gaboon, where the mean annual temperature is 95° F., the rainfall from 100 to 110 inches, the relative humidity very high, rarely under 90. The heaviest rains occur between September and January.

The next district to be mentioned is that at the mouth of the Congo. The mean annual temperature is 76° F., the annual range 9° F.; the rainfall 43 inches; the relative humidity for the year 75 per cent., the annual range 13. The rainy season is from November to May, with prevailing westerly winds.

With regard to the next region, Angola, the following data represent the climate at St Paul de Loanda. The mean annual temperature 74° F., the annual range 12° F.; the rainfall 13 inches; the relative humidity 82 per cent., and the annual range of humidity 10. Here there are two rainy seasons—October to December, and March to May.

In the districts to the south, namely, Benguela and Mosamedes, the country is drier, the rainfall varying from 10 to 30 inches, the mean annual temperature being about 68° F., the annual range 20° F.

The disorders met with on the west coast of Africa, and which may be regarded as the most prevalent tropical diseases, are numerous, and we have to deal with two of them which are unknown on the eastern side of the continent, namely, yellow fever and yaws. Following the rough order which I have hitherto adopted, we find that typhoid fever is occasionally met with in Senegambia, more rarely, perhaps, in Sierra Leone. It is not reported from the Gold Coast, but it occurs in the Niger district, as also in Gaboon. There are no reports of typhoid fever from the Congo, but, as it is stated that typho-malarial fever occurs there, in all probability it does exist; for as I shall point out subsequently, I am convinced that typho-malarial fever does not exist as a disease *per se*, but that in all cases where typho-malarial fever is reported we are really dealing with patients in whom both typhoid fever and malaria are present.

Diarrhœa and dysentery are extremely common in Senegambia, being most frequently seen from September to November; also prevalent to a slightly less extent in December, January, and February. Both diseases occur also in endemic form in Sierra Leone; they are however not quite so frequently met with as elsewhere. Diarrhœa is often met with on the Gold Coast, but dysentery is more rarely seen, at any rate among the Europeans. Both diseases occur at Lagos, as also with extreme frequency in the Niger district. Diarrhœa is very common at Gaboon, but dysentery, although endemic, affects the natives chiefly. Probably the diminution of cases amongst the white residents is due to sanitary precautions. On the Congo coast diarrhœa prevails, as also dysentery, but this disease is not nearly so frequent as in the Upper Congo district, to which reference will be made later on. In Angola, diarrhœa is fairly common, but dysentery is more rarely seen, and when it occurs it is chiefly after the rains.

Severe epidemics of dengue occur in Senegambia, but curiously enough that is the only place on the west coast of Africa where this disease obtains. Cholera visited Senegambia in 1868, but has not been recorded as occurring elsewhere on the west coast of the continent, although we do find scattered notes referring to sporadic cholera on the northern part of the west seaboard. Now in 1893 cholera is again ravaging Senegambia.

Beri-beri is very prevalent in the Congo district near the coast, as also is a disease called the sleeping-sickness or negro lethargy, a disorder which has a rather wider distribution, namely from Senegambia to this district. It is at present a moot question whether these two diseases are identical. It is also thought by some that both diseases are really due to the *Anchylostoma duodenale*, and if so they would be identical with Egyptian chlorosis, which is undoubtedly due to the presence of that worm.

Yaws or frambœsia is very frequently met with on the west coast of Africa, from Senegambia on the north as far south as Angola.

The area of distribution of yellow fever is at present limited in Africa to the west coast from 19° N. to a point on the mainland opposite Fernando Po.

Hirsch gives the following chronological survey of yellow fever epidemics on the west coast of Africa from 1816 :—

1816. Sierra Leone, Congo coast.	1859. Sierra Leone, Senegambia.
1823. Sierra Leone.	1860. Gambia and the Congo coast as far as Angola.
1825. Sierra Leone.	1862. Gambia and the Congo coast, Gold Coast, Benin coast.
1829-30. Sierra Leone.	
1830. Senegambia.	
1837-39. Sierra Leone.	1864. Sierra Leone.
1837. Senegambia.	1865. Sierra Leone, Congo coast.
1845-47. Sierra Leone.	1866. Sierra Leone, Senegambia.
1852. Gold Coast.	1867. Senegambia.
1857. Gold Coast.	1868. Sierra Leone, Senegambia.
1858. Senegambia.	1878. Sierra Leone (?), Senegambia.

Diphtheria and scarlet fever are unknown, but measles occurs from Senegambia to Angola. Diseases of the chest, such as bronchitis, pleurisy, and pneumonia, are met with all along the coast, but phthisis is extremely rare among the natives, although it is of course sometimes seen amongst the whites, who have in all probability contracted the disease before proceeding to the coast. Leprosy is frequently met with both in its anæsthetic and tubercular forms in the native population on the whole of the West African coast. Syphilis and rheumatism, especially the acute form, are also extremely prevalent.

Elephantiasis arabum also occurs with comparative frequency over the whole coast, being especially prevalent on the Gold Coast, where also ainhum is frequently seen.

Small-pox occurs everywhere, and very severe epidemics

often spread with extremely fatal results amongst the native population. Very few cases of cancer are met with on the west coast of Africa, but tropical ulcers are extremely common.

With regard to hepatitis and tropical abscess of the liver, there is considerable variation in the occurrence of these diseases, both amongst white men and natives. For instance, hepatitis is extremely prevalent amongst the whites in Senegambia, but is not often seen amongst the natives. It is comparatively rare in the Niger district, but is more frequently met with on the Congo coast. On the whole, tropical abscess of the liver is less frequently seen than one would expect, and it would appear to be a sequel to severe attacks of dysentery. This statement, however, should be taken with caution, as if it is true it is very different from what usually obtains in India and other tropical countries. Whooping-cough is said to occur in Senegambia, but I have seen no account of it elsewhere.

Amongst the native population over the whole west coast of Africa, skin diseases are common. Insanity is not frequent, but tetanus often occurs, especially amongst children. Rickets are comparatively rarely seen.

Anæmia is invariably found in Europeans throughout the west coast of Africa, and is usually due to malaria; it may however be due simply to residence upon the coast. Guinea-worm is especially frequent between Senegambia and Cape Lopez. In Senegambia it is met with not only on the coast, but in the more elevated region which extends from Bakel to Galem, though the parasite does not infest the banks of the Casamauce. The Sierra Leone coast is less extensively infected by the guinea-worm than the Grain Coast, Ivory Coast, Gold and Slave Coasts. The worm is met with throughout the Niger district, and also in Gaboon.

It is to be noticed that on these coasts various places, such as Cape Coast Castle, Elmina, Cormantia, and Accra, are especially affected, whereas the surrounding country very often is free from the parasite. I do not believe that there is any connection between guinea-worm and *Elephantiasis arabum*.

Amongst the natives, nervous diseases are rather frequently seen. Hemiplegia, paraplegia, epilepsy (especially in women), acute mania, and dementia are met with.

Throughout the whole coast endemic dropsy occurs, but here again, as I said before in referring to the east coast, I think this is a form of beri-beri. Ague-cake is very common, especially in Creoles. It is said, however, to be uncommon in Sierra Leone, and Gore attributes its rarity in that region to the ferruginous character of the soil. Goitre is not frequently seen on the west coast of Africa; there are, however, some reports of its occurrence in the Cameroon district.

The west coast of Africa presents four seasons, which generally begin and end as follows, allowing for latitude and local peculiarities. The summer season extends from February 15 to May 15, the rainy season from May 15 to August 31, the harvest season from September 1 to November 15, while the harmattan or cold season begins on November 15 and ends on February 15.

The summer is hotter than any of the other seasons, the heat being greatest in the region of the trade winds, and greater in Sierra Leone than on the Gold Coast and Bight of Benin. In addition to the prevailing winds, which vary from N. and N.E. in February to S. and S.W. in April, the Gambia region is subject to the simoon, a hot wind charged with fine sand which blows from the desert, destroying vegetation and causing much distress, especially to those suffering from respiratory diseases. Whirlwinds are also prevalent during this season.

The rainy season is ushered in by a cloudy atmosphere and frequent tornadoes. The rain gradually increases till in July and August it descends in torrents, in many places inundating the country and washing away huts and bridges. In the Gambia region the commencement of the rains is marked by occasional dirt gales, a strong wind carrying the surface earth along with it. The actual rainfall varies in different places, and also in the same place from year to year. It is greatest during the night, and at the beginning and end of the season the fall is limited to the night time. In the rainy season sand-flies and mosquitoes are especially troublesome.

The harvest season or autumn is the most unhealthy part of the year in West Africa. The rains moderate, the swamps begin to dry, while the decomposing vegetable matter favours the spread of malaria. The heat, though not actually so great as in summer, is far more oppressive from the moisture of the atmosphere. The south-west monsoon is the prevailing wind, and at the end of the season the north-east monsoon, while beyond the region of the "trades" there are the usual land and sea breezes. At the termination of the harvest the electrical condition of the atmosphere is much disturbed, this being followed by the cessation of rain for some months.

The harmattan or cold season is so called from a wind of very peculiar character which occurs at this time. It is cold and extremely dry, owing to its course from the east over the Sahara. On its approach, vegetation of every kind is shrivelled up, and the lips and eyes of those exposed to it suffer from its parching effect. It is accompanied by a thick fog and mist, composed of particles of fine sand of the desert. It only blows for a few days in the season, the prevailing wind being from the south-east. This is the dryest season of the year, and is generally healthy; vegetable matter being shrivelled instead of decomposed, as during the autumn, does not favour the production of malaria.

The months of February, March, April, October, and November may be regarded as the hot season of West Africa. During this period the increased heat gives rise to certain physiological effects on the human body. The temperature rises to 100° F. or even 102° F., the pulse is accelerated, the respirations diminish in depth and frequency, the urine becomes concentrated, and activity of the skin is enormously increased.

Although in the summer season the heat is most intense, it is not so unhealthy as the autumn, when the heat is combined with moisture. (The month of October is specially unhealthy.) The heat at first causes an exaltation of general sensibility, but afterwards this gives way to marked depression.

Exposure, intemperance, bodily or mental depression, may result in ardent fever, with cerebral or hepatic complications. Dysentery and diarrhœa occur chiefly when the water-supply is impure from storage or contamination with decaying organic matter.

The chief diseases met with during the hot season are ophthalmia, dysentery, intermittent fever, rheumatism, leprosy, guinea-worm, and prickly heat.

The rainy season is far more unhealthy than the hot months, especially at its commencement, and to a less extent at its close. The temperature of the air falls at the beginning of the rains, producing a feeling of vigour after the intense heat; but the moist atmosphere, combined with the sun's rays, greatly increases the perspiration, and along with this there is relaxation of the muscular system, with cardiac debility and congestion of the internal organs. Should the rainfall be scanty, irregular, and alternating with hot weather, severe outbreaks of fever are to be expected from the formation of pools of stagnant water rich in organic matter. Dysentery and diarrhœa are also prevalent from the contamination of the water-supply; while in addition to the diseases mentioned above, diseases of the respiratory system are also met with.

It is likewise to be noted that though guinea-worm may occur at any season of the year, it seems to be more troublesome during the colder months. Elephantiasis and goitre also seem to commence more frequently in the cold season, though afterwards the seasonal influence on their progress is less marked.

The season of the harmattan is the most healthy part of the year, the drying action of the wind stopping the decomposition of vegetable matter, and hence the production of malaria. When it begins, patients suffering from malaria rapidly become convalescent, while other diseases are frequently benefited in like manner. The dry, cold air, while it braces up the body generally, reddens the skin and renders it dry and harsh. The nostrils and pharynx become dry, the lips chap, and the eyes may become inflamed. The sensible perspiration is almost arrested, and the activity of the kidneys is correspondingly increased. In the absence of free renal secretion, copious diarrhœa may ensue, and may persist in spite of treatment till the wind abates. The other diseases are chiefly the result of the internal congestion set up by the cold, congestion going on to sub-acute inflammation of the liver and spleen, hæmorrhages from mucous membranes, and abortion. Infants suffer severely from cold in the intervals between the harmattan. Slight attacks of fever unattended by prostration may occur; also rheumatic attacks, but these are arrested on the recurrence of the desert wind.

According to Horton, the harmattan has a decided effect upon small-pox; the pustules soon heal up, and the disease disappears. Persons vaccinated whilst the harmattan wind is blowing are not protected against small-pox, as the vaccine will not take.

I shall now deal generally with the incidence of malaria on the west coast of Africa. All I can do is to give a summary, which I hope may indicate, with some attempt at accuracy, not only where the disease is prevalent, but where it is met with in its most grave forms.

Some amount of confusion obtains as to the distribution of the graver forms of malarial fever in West Africa. This is due to the fact that some observers hold bilious remittent fever, blackwater fever, endemic hæmaturia, and typho-malarial fever to be distinct diseases. I do not agree with this view, and in what follows it must be distinctly understood that I entertain the conviction that these so-called various diseases are simply malarial fever distinguished by some prominent symptom which has given the name to a variety of that fever, as indicated above. Believing as I do that malaria is a disease *sui generis*, caused by the hæmatozoon discovered by Laveran, I classify the results of its action on human beings as follows:—Intermittent fever of varying types — *i.e.*, quotidian, tertian, quartan ague; remittent malarial fever, including bilious remittent fever, blackwater fever, hæmoglobinuric fever, and endemic hæmaturia; pernicious malarial fevers, including the comatose and algid varieties; masked malaria, including brow ague and the various neuralgias connected with malaria, and finally malarial cachexia.

To the causes of malaria, and its production being favoured by local circumstances, or prevented in some cases, I shall not now allude, preferring to deal with such matters later on, when speaking of the prevention and cure of the disease.

Although, as before mentioned, the West African coast is called "the white man's grave," yet it is undoubtedly true that to a certain extent it is not malaria which causes, or perhaps one should say has caused, the very high death-rate of even 50 per cent. amongst the whites on the coast. In the past, at any rate, this death-rate has been due to the fact of diseased individuals proceeding to Africa, to want of knowledge of the precautions necessary for a residence there, and unfortunately, in many cases, to the wilful ignoring of prophylactic measures and of a well-ordered life.

In Senegambia malarial fever causes at least 40 per cent. of all the cases of disease. In some places the admissions into hospital from malarial fever rise as high as 70 to 80 per cent. The greater number of cases occurs during the rainy season, or between June and November, and chiefly, in all likelihood, at the commencement and at the end of the rains, and probably the pernicious, and bilious and hæmaturic fevers happen chiefly during the rainy season, when the mean monthly temperature is highest. With regard to the hæmaturic fever, it probably occurs only under certain circumstances, namely, in debilitated individuals, after extreme strain or excess, or after a severe wetting following repeated attacks of ordinary intermittent fever. The bilious remittent fever may be considered as an acclimatising fever, occurring chiefly in newcomers.

Next, in Sierra Leone the amount of malaria is extreme. The type of fever is here very severe, a very fatal remittent type being most commonly met with in the whites. Again we find that here also it is during the rainy season that malaria is most virulent. On the Gold Coast the same conditions exist, and grave remittent fevers, with bilious and hæmaturic symptoms, are extremely common, as also are pernicious fevers.

Along the rest of the coast malaria is very virulent as far as Cape Lopez. Some idea of its prevalence will be gained by mentioning the death-rate from 1878 to 1888. Unfortunately, the cause of death is not stated in the reports, but malaria appears to have been the chief cause.

There are no means of ascertaining the number of Europeans in Gambia during the years in question, but I have reason to believe there were about 54. The average death-rate there was 10·10 per cent. At Lagos, from 1879 to 1888, there was an average of 110 European residents; and the average death-rate was 10 per cent. At Sierra Leone, the population during the same years was about 271, of whom 108 were a floating population, *i.e.*, belonging to ships in the harbour. The total deaths of the resident population were 44, of the floating population 31, giving a total number of deaths during the ten years of 75, being 42 per cent. On the Gold Coast during the same year there was an average of 66 Government officials. There were 34 deaths, the average death-rate being 51·43 per thousand. The non-official population was about 126, the number of deaths 106, the death-rate being 81·48 per thousand; or, taking the officials and non-officials together, the death-rate per thousand would be 68·08, the average European population being 192.

In the Niger district, remittent fever is very virulent, and this district is remarkable for the prevalence of pernicious malarial fever. It is here also that we have most reported cases of so-called typho-malarial fever, the only other region where this complication seems to prevail being the Congo district; at Gaboon and in its near proximity malaria is most rife during the first three months of the year.

In the Congo coast region the average death-rate from malarial fever appears to be about 30 per cent. I shall refer to its prevalence in the upper regions of the Congo subsequently.

To the south of Loanda, where the fever is very rife, malaria decreases in intensity and importance as we proceed southward.

V. THE SAHARA AND SOUDAN.

This area, which includes the country as far south as 10° N. latitude, takes in the districts of Bornu, Wadai, Darfur, and Kordofan, and must be dismissed in a few lines, statistics being wanting with respect to it. On the whole the climate is intensely hot in summer, but the diurnal variation is very great. Sometimes the thermometer is many degrees below freezing point at night, although during the day it may be as much as 115° F. In the northern part of this district the mean annual variation or range of temperature is from 40° to 50°. In the southern part, in Wadai, Darfur, Kordofan, the range is considerably less, being only from 10° to 20° F. Over the greater part of the area the mean annual temperature is over 80° F., although in the district round Wadai it varies from 70° to 80°, this being due to the altitude of the country, which is from 3000 to 8000 feet. On the whole, this district may be said to be healthy. Except in the oases, malaria is almost absent. In Darfur and Kordofan dysentery, diarrhœa, and typhoid fever are met with; syphilis and bronchitis are fairly common; and in the southern part of Darfur the guinea-worm is very frequent, the area of its northern distribution at this point being about 11° N. lat. Goitre is not infrequently seen in inhabitants of the Djebel Marra. The most malarious parts of this district are the low swampy regions round Lake Chad, Dibbe, and Filter, and the province of Fezzan.

VI. EQUATORIAL CENTRAL AFRICA,

including the Congo Free State, and reaching south to the 18th degree of S. lat.

The mean annual temperature of the whole of this region is about 78° F., being slightly lower to the west of the Victoria Nyanza in the Ruwenzori mountain region, on the high plateau to the east of the Victoria Nyanza, and in Msiri's kingdom, where it varies from 65° to 70° F. The mean annual range of temperature in the Congo region is less than 5° F.; in the remaining area it is between 5° and 10° F., except in the region to the south of the Victoria

Nyanza, bounded between 32° and 38° E. long., where the mean variation is from 10° to 20° F. The mean annual rainfall is about 50 inches, except in the centre of this area, where it reaches about 100 inches, this including the Congo forest district and the country to the west of Tanganyika as far south as 10° S. lat. In this district also the relative humidity is over 70 per cent., but over the whole district the relative humidity may be taken as about 66 to 68 per cent.

The whole region is highly malarious, but I will give a brief account of malaria as I met with it in Central Africa. The area referred to extends from Khartoum in the north as far south as the Victoria and Albert Lakes, and also includes the Rohl and Bahr-el-Ghazel districts. The distribution of malaria in this region is unequal, and the topography of the country exerts an influence both upon its frequency and its severity. In low-lying swampy regions malaria is very common, the natives suffering to a considerable extent from mild attacks of intermittent fever. Occasionally one sees a case of well-marked remittent, but perhaps the most frequent variety met with is a form which at first sight appears to be a continued fever lasting from five to seven days. But when these fevers are examined carefully, it is found that they are really either mild remittent (for there are distinct remissions, which, however, must be carefully looked for) or they are quotidian, with badly-marked paroxysms. A very brief cold stage occurs daily, then follow eighteen or twenty hours of hot stage with a temperature of 102°-103° F., followed by an hour or so of apyrexia after a very slight perspiration. Throughout the whole region the natives, who are fairly stationary in one district, suffer comparatively little, and from but slight attacks of intermittent fever, but if removed to a new locality they suffer from much more severe attacks, from remittent fever for the most part, and it is a noteworthy fact that after slave raiding or war, numbers of men are struck down by severe forms of fever.

In the higher regions the fevers become more rare until one reaches districts having an altitude of from 3000 to 4000 feet, where they almost entirely disappear. A good example of this may be seen in the country to the north-west of the

Albert Nyanza, also in Central Unyoro, in the Shuli district, to the south-east of Dufli, and in Uganda, in the Kahura district.

With regard to the effect of malaria upon Europeans and Egyptians, one notices marked differences; and here the personal equation comes notably into play. Some suffer very little from malaria, others suffer from severe remittents and from what some call bilious remittents and blackwater fever. On the whole, Europeans suffer less from anything more than an occasional attack of typical intermittent fever than do the Egyptians. Inactivity, severe marches in the sun or by moonlight, and fatigue, are the predisposing causes of attacks of fever in Europeans. In all the old Egyptian stations which I have visited, the Egyptian troops and officials suffered excessively from malaria. It was very fatal, or if not fatal, it induced such marked debility that they were greatly incapacitated for ordinary employment. I found that the "spleen test" was very useful in ascertaining approximately the salubrity of a district. The Bahr-el-Ghazel district is, owing to its very abundant water-supply and its many swampy areas, excessively malarious. The country to the north of the Bahr-el-Arab is comparatively exempt.

I pass on now to deal briefly with the subject of enteric fever. At Khartoum it is endemic, and no wonder, when one considers the filthy condition of, and want of all sanitary precautions in that town, and the quagmire into which it is yearly transformed at high Nile. After the inundation of the Nile the disease spreads all over the so-called island of Meroë. In all the districts I have mentioned above, as also in Kordofan, I met with cases of enteric fever, but they varied in frequency, not so much with the character of the country or the climatology, as with the habits and customs of the natives, and their sanitary surroundings. The disease was most frequently seen in the larger settlements in the Bahr-el-Ghazel districts. There, where the slave-dealers were in the habit of crowding together thousands of slaves, the filthy condition of the places can be well imagined, and it was in these hotbeds of disease that I saw most cases of enteric fever. Still, I met with the disease at Bohr on the White Nile, at Foweira, at Magungo, just to the north of the Albert Nyanza, and I witnessed one epidemic in Uganda. I say epidemic, because it was curious to notice that, generally speaking, enteric fever seemed to stop short directly an area was reached in which the banana forms the staple food of the population. It was far more frequently met with in those districts where the people lived chiefly upon grain.

I was surprised in my journey in Central Africa to notice the distribution of phthisis, for, although bronchitis, pleurisy, and pneumonia were constantly seen in nearly all the districts through which I passed, the cases of phthisis which I was able to observe were few and far between, and corresponded in a marked manner with the absence of malaria, at any rate in its most intense forms. From Khartoum along the valley of the Nile as far as the Albert Lake, through the swampy districts of Unyoro and Uganda, I can recall having seen very few cases of phthisis (in Uganda some eighteen or twenty). Subsequently, however, I saw a considerable number of cases in the Shuli district, at an altitude of 3000 to 4000 feet, where malaria is very rare. Again, in travelling through the Bahr-el-Ghazel district I saw a considerable number of phthisical individuals, not inhabitants of that province, but men or women, soldiers or slaves, who had come from the elevated districts in the Monbuttu country. Farther north, at Dara, I again met with phthisis in people who inhabited the highlands of the Djebel Marra district, where I was informed that malarial fevers were entirely absent.

With regard to the other diseases of Equatorial Africa, what follows refers to the districts mentioned above, and with which I am personally acquainted; but from what I have read on the subject, I have reason to believe that the same diseases obtain to the south.

Small-pox occurs in epidemics, and is very fatal. Measles and scarlet fever are unknown. Rheumatism is common everywhere, and cholera has on various occasions passed through the country in the form of epidemics. With regard to phthisis, it is very rarely met with throughout the whole of the northern part of this area, except in the high region to the north-east of the Victoria Nyanza, where it is more common. Diseases of the chest are found throughout the whole region, but bronchitis is far more common than either pleurisy or pneumonia. A form of plague has visited Uganda on several occasions, and there are reports of it having occurred on the White Nile to the south of Lado and in the Bahr-el-Ghazel district. Guinea-worm is most prevalent to the west of the Nile throughout the province of Rohl and Bahr-el-Ghazel. Round worms are also met with, and Yaws is occasionally seen. Syphilis is widely spread in those districts where the slave trade has been carried on, but it is not prevalent in other regions. Leprosy is met with, but not extensively. *Elephantiasis arabum* occurs, especially on the west of the Nile to the south of Lado. Skin diseases are extremely frequent, except in Uganda, and boils are of common occurrence everywhere.

With regard to nervous diseases, temporary insanity is often met with, but it is rare to see cases of permanent aberration. Epilepsy is fairly common, and occurs chiefly in girls.

Ophthalmia is comparatively frequent, although it is not nearly so prevalent as in Egypt. With regard to diarrhœa and dysentery, both diseases are met with throughout this region. They are more prevalent throughout the Nile valley, and in the west, than in Uganda and Unyoro. The so-called blackwater fever certainly occurs, and so does typho-malarial fever, but in all cases I came to the conclusion that these were different varieties of remittent fever.

Passing now to the Congo region of Equatorial Africa, we find that malaria is prevalent over the whole of the district. All varieties occur, from mild attacks of intermittent fever lasting three days, to the most pernicious forms of fever, such as are seen at Vivi and Stanley Pool. It is said that the mortality of Europeans in the Central Congo region is about 25 per cent. Here, too, the so-called blackwater fever is common. There are no reports of enteric fever from the Congo, but typho-malarial fever is reported. I believe, however, that it is really simple severe remittent fever with typhoid symptoms. Dysentery is very common, and so is hepatitis, and tropical abscess of the liver is also met with. Phthisis would appear to be rarely seen, if at all. The ordinary diseases of the chest, however, occur; as also do rheumatism and Egyptian chlorosis. Leprosy and yaws are both endemic in the Congo region, as also the disease of sleeping-sickness, as it is termed by writers from this region. As previously mentioned, I regard this disease as beri-beri.

We next come to the remaining part of Equatorial Africa, that to the south of the Victoria Nyanza, surrounding Lakes Tanganyika and Nyassa, as far south as the Matabele country. Throughout this region malaria is again extensively met with, except in the highest regions. Enteric fever exists to the south of the Victoria Nyanza, and it appears to occur occasionally as far south as the Zambesi. Both diarrhœa and dysentery are found, but dysentery is reported to be exceedingly severe throughout the whole district, far more frequent in fact than it is either to the north or west. Rheumatism, dengue, leprosy, and syphilis are all prevalent, and so are the ordinary respiratory diseases, but phthisis is extremely rare. *Elephantiasis arabum* is met with to the south-west of Tanganyika especially, but isolated cases are occasionally seen throughout the whole district. Tropical ulcers are seen very frequently in this district; so is ophthalmia, and probably beri-beri is widely distributed. Curiously enough, diseases of the liver and tropical abscess are rarely seen here. Diphtheria certainly exists, but it is not very prevalent.

VII. SOUTH AFRICA,

including the country south of 18° S. lat.

The area with which we have now to deal has a climate which differs much from any we have hitherto considered; it resembles more nearly that of the temperate zone, in which zone indeed the greater part of the district lies. The country varies considerably in altitude, lying generally from 600 to 10,000 feet above sea-level.

The mean annual temperature varies from 64° to 72° F.,

excepting in the Kalahari Desert, where it rises to about 76° F. The mean annual range of temperature varies somewhat: on the coast it is from 10° to 20°; inland it is from 20° to 40° F. The annual rainfall varies considerably. It is least on the west, where it is under 10 inches; indeed, in the north-west, it is only about 3 inches annually. On the eastern coast the precipitation is heavy, varying from 18 to 40 inches. Between these two areas, in the South African Republic and Orange Free States, the annual rainfall is from 10 to 25 inches.

In the most northern part of this area we have two districts—Mashonaland and Matabeleland. The altitude of this country is about 4000 feet. The rainy season is from November to February, and, on the whole, the district appears to be fairly healthy, although we have not sufficient reliable information upon which to offer a final opinion. We know, however, that dysentery, diarrhœa, and rheumatism are very prevalent, and that malaria also prevails, though, perhaps, not so extensively as it does to the north of the Zambesi,—*e.g.*, Bechuanaland, which has the same elevation. In the west we have the Kalahari Desert, with a very slight rainfall and with considerable heat; but the eastern district is better known, and therefore, probably, we have information of more numerous diseases, although, on the whole, the country would appear to be healthy. The rainy season is from December to April; the average rainfall is 25 inches; the temperature may be as high as 85° or 90° F. The climate is, on the whole, dry and invigorating. Probably the most fatal disease in this district is dysentery. Enteric fever is also met with; malarial fevers are unimportant; measles and small-pox occur in epidemics; phthisis is unknown; bronchitis, pleurisy, and pneumonia are only very rarely seen. Syphilis is now common; rheumatism and hepatitis are both met with; ophthalmia is extremely common, as also is whooping-cough. Leprosy is not endemic; this is rather surprising, as the disease is prevalent at the Cape.

The Orange Free State is an elevated plateau lying 4000 to 5000 feet above sea-level. It possesses a remarkably dry climate, the humidity being about 55 per cent. During the six hottest months of the year the average maximum temperature is 82° F., the average minimum temperature 55° F. The dust-storms are the only drawback to this climate. The average maximum temperature for the six coldest months is 66° F. The rainfall is 16 inches. Diseases of the chest are very rare in this district, and it forms an admirable health resort for phthisical patients. Dysentery and diarrhœa are the chief diseases to be dreaded here.

The Transvaal has the same altitude as the Orange Free State, and possesses groups of mountains. It also has a dry climate, which is salubrious and exhilarating. The rainy season lasts from October to March, with an average rainfall of 30·74 inches. The principal diseases in this district are typhoid fever, dysentery, and diarrhœa. Leprosy and diphtheria are also met with; the occurrence of malaria is unimportant. Owing to sudden changes in the temperature, bronchial catarrh is fairly common. Phthisis is not seen.

In Natal, including Zululand, Basutoland, Griqualand East, and Pondoland, the climate in this region varies considerably. At Durban, on the coast, the mean annual temperature is 77° F., the mean range 18° F., the rainfall is 33 inches. At Pietermaritzburg and the interior, the mean annual temperature is 68° F., but the sudden changes from hot, dry land winds to moist sea breezes are trying. July to September are the most trying months in Natal. The rainfall at Pietermaritzburg is 31·87 inches. With the exception of isolated areas at the coast, and in some of the gullies in Zululand, malaria is practically unknown, but typhoid fever and dysentery are fairly common, as also is diarrhœa in the hot season. Respiratory diseases are very rare, but diphtheria, small-pox, measles, and scarlet fever sometimes occur in epidemics. Phthisis, except in imported cases, is practically unknown.

Cape Colony is, on the whole, very healthy. Temperature and rainfall vary in different places. At Cape Town itself the mean annual temperature is 67° F., the annual range

38° F., the rainfall 23·12 inches. At Port Elizabeth the rainfall is only 19·71 inches, and at King William's Town, 16·48 inches. At Graham's Town the air is bright and exhilarating; the mean annual temperature is 60° F., mean annual range 15° F., rainfall 22 inches, and, occurring as it does chiefly in summer, it keeps down the temperature and secures remarkable equilibrity. In this region malaria may be said to be absent. Again here, typhoid fever, dysentery, and diarrhœa are the most frequent and fatal diseases, and rheumatism, also, is very prevalent. Syphilis, leprosy, and scrofula are widely prevalent, but respiratory diseases are unimportant. Pneumonia, however, is more frequently seen in Cape Colony than it is farther north. Phthisis, owing to the number of imported cases, is more prevalent than one would expect, but apart from this, it is probable that the disease does exist in the Colony more than in other parts of the district we have just had under review. Scarlet fever, diphtheria, small-pox, and measles occur in epidemics, but infrequently. Heart disease appears to be specially prevalent. Cholera has never visited the Cape, and hydrophobia is unknown. Diseases of a parasitic nature are rare, and hydatids are infrequent.

VIII. AFRICAN ISLANDS.

Madagascar, the largest African island, is situated in the Indian Ocean, between 12° and 25° S. latitude. It can be divided into a low coast-line, and extensive highlands having an altitude of from 3000 to 4000 feet. The mean annual temperature of the coastal region is between 74° and 80° F.; the mean annual temperature of the highlands is 64° to 70° F., the mean variation between 5° and 10° F. The annual rainfall on the western half of Madagascar varies from 25 to 50 inches; on the eastern half it varies from 50 to 100 inches. At Nossi Bé, to the north-west of the island, it is over 100 inches. Malaria is most prevalent during the first three months of the year, and it prevails over practically the whole of the island. It is especially prevalent on the coast, and least so in the central provinces. Typhoid fever is also prevalent in the centre of the island ;· dysentery, on the other hand, occurs more frequently on the coast. Diarrhœa is met with all over the island; phthisis is said to occur in the higher regions, but, with the exception of pneumonia, diseases of the chest are comparatively rare. Rheumatism, leprosy, and syphilis are common. Beri-beri occurs in epidemics; diseases of the liver are fairly common.

The Seychelle Islands, situated between 3° 30' and 5° 30' S. latitude, are on the whole healthy. The mean annual temperature is about 77° F., the mean annual variation about 15° F., and the mean annual rainfall about 80 inches. Malaria is practically unknown, and diseases of the chest, with the exception of phthisis, are rare. The chief diseases of these islands are dysentery, phthisis, and affections of the liver. Leprosy and syphilis are fairly common.

Mauritius, an island lying between 20° and 20° 30' S. latitude, is hilly, and has an elevation of from 500 to 700 feet. The mean annual temperature at St Louis is 78° F. April to November is the coolest part of the year. The rainfall is 70 inches. Up to the year 1866 the island was free from malaria, but since then it has been very malarious, both intermittent and remittent fevers being exceedingly frequent. The so-called bilious remittent fever is very common, and typhoid fever is also frequently met with. Dengue is epidemic; dysentery and diarrhœa are common, so are leprosy and syphilis; probably also beri-beri. *Elephantiasis arabum* is endemic. Chest affections are comparatively rare; but phthisis is sometimes seen. Ophthalmia is exceedingly prevalent, and hepatitis seems general amongst the white population.

The foregoing summary of the climatology of the various artificial regions into which I have divided Africa, and the account of the distribution of diseases occurring in them, has necessarily been somewhat dry and tedious; indeed, it has been very difficult to compress the necessary information within at all reasonable limits. I have now to give a general statement as to the nature of the prominent diseases met with in Africa, their cause and prevention, and the methods of treatment which are employed for their cure. It would of course require a volume to deal with the subject adequately, but I hope I shall be able to say sufficient, to give an intelligent outline of the subject.

It will be well in the first place, to describe what methods are employed by the natives in Africa in combatting disease, in so far as any obtain, and then to describe the methods of treating disease which are indicated by modern medicine.

It would be a hopeless task were I to attempt to describe in detail the minutiæ of medical and surgical treatment adopted by the natives for accident and disease in various parts of Africa, nor do I consider it necessary. It will be better, I fancy, to give a more general outline of the subject, but it is very necessary to state that I am doing so, for different methods obtain among different tribes, and even neighbouring tribes may have different ideas and customs with regard to any single disease. Were I not to state this definitely, I should easily lay myself open to criticism by any observer who had a knowledge of one single tribe exclusively.

The natives of Africa, and by these I mean the Negroes, Bantus, and the Arabs inhabiting the Soudan, excluding the Egyptians and the inhabitants along the northern African coast, have many superstitions with regard to medicine. Broadly speaking, one may say that the questions of life and death, health, and sickness, or accident, bulk largely in their ethical cogitations, and I do not think I am going too far when I say that the natives consider that all the evils which flesh is heir to result from the malign influence of the powers of the air. So they have pictured to themselves gods of small-pox or famine, gods of thunder, in fact a hierarchy of spirits who shape the destinies of men, and who may torment them if not controlled by charms or propitiated by votive offerings. Recognising this fact, remembering the native suspicious nature, their belief in fetishes and charms, the evil eye, etc., it is not to be wondered at, that incantations play a marked rôle in their treatment of disease; but it is a mistaken notion to suppose that there is not a basis of true common sense underlying the hocus-pocus of which one hears so much in travellers' tales. The natives are observant; they recognise the fact that such diseases as small-pox and syphilis are contagious, and they attempt with varying success to prevent their spread. In many districts in Africa a rigid system of isolation is practised with regard to small-pox. The patients are treated in a hut set apart for the purpose, they are attended only by those who have previously suffered from the disease; a definite dietary is prescribed for them, simples are administered, the pustules are pricked with a sharp thorn, and thereafter various unguents, famed for their healing properties, are applied to the patient. In one district, at any rate, in Africa, inoculation is practised with the syphilitic virus, and neither young man nor maid there may marry until they have been through the inoculation ceremonies. The African natives have a considerable knowledge of the virtues of plants. Sudorifics, diuretics, febrifuges, purgatives, and emetics are known. Numerous barks and plants are selected as possessing these various properties, and, accompanied by incantations, are applied with no little success. The actual cautery and cupping are well-recognised procedures, and are employed in many diseases.

With regard to surgical operations, amputations are practised in some parts, but in many the patients would sooner die than suffer mutilation. Splints for the treatment of fractures are known and widely utilised; hæmorrhage is stayed in various ways, either by cautery, by the application of boiling oil or water, or of compresses composed either of astringent herbs, coffee grounds (in the Soudan), or cobwebs. It is when dealing with such diseases as epilepsy, insanity, or nervous disorders, that the natives completely fail, and where incantations pure and simple are resorted to. In dealing with labours, considerable ingenuity is often manifested; turning is sometimes practised; the child may be expressed, so may a retained placenta; in difficult labours positional treatment is in vogue in many places, and in some, even abdominal section is practised with more or less success.

As is well known, poisons and their antidotes find a not

unimportant place in the medicinal lore of Africa. At least one of their poisons—strophanthus—is now a well-recognised medicine for heart disease, which has been adopted all over the civilised world.

What medical and surgical knowledge the natives possess is, as has been indicated, bound up with witchcraft, and the knowledge is usually handed down in families; or in some cases, as for instance in regard to poisons, it is regarded as a tribal secret, and the method of preparation of poisons and their antidotes is only communicated to such as, after initiation, have proved themselves worthy to be the recipients of that knowledge.

Such then in brief is the character of the native methods of treating disease. It now remains for me to indicate what light European medicine throws upon the treatment of disease in Africa, and how far modern medicine can either prevent, diminish, or cure the numerous maladies to which white races, as has been seen, are liable when transplanted to this foreign soil.

Certain characteristics should be possessed by individuals who leave a temperate climate to reside in Tropical Africa. First with regard to temperament. As was pointed out by Dr Moore, the *sanguine* temperament is associated with a tendency to congestive affections, to a rapid and irregular development of disease, to head affections, abscess of the liver, and scurvy.

Persons possessing this temperament are characterised by active muscular systems, and high animal courage, but they live at high pressure, and cannot sustain slight exposure to noxious surrounding influences. The *nervous* temperament is very sensitive, but there is much energy and capacity for endurance of fatigue, privation, and exposure; persons, however, possessing this nervous temperament, are prone to diseases of the nervous system and hepatic affections. With regard to those of *bilious* temperament, the frame is powerful and the person possesses great endurance. He has the least sensibility of all to morbid disturbances and external impressions. He has no extraordinary tendency to liver affections, except in the extreme form, the melancholic bilious temperament. "The bilious temperament possesses the good qualities of the nervous, without its irritability, and of the sanguine, without its susceptibility to external impressions." Individuals possessing the *lymphatic* temperament do not resist disease well, and readily suffer from disease of the liver and derangements of digestion. It follows, therefore, that the bilious, or bilio-nervous temperaments are best fitted for residence in the Tropics; then persons of the sanguine temperament, but those would only stand the climate for a short time. Persons having a lymphatic temperament should stop at home. No one possessing a syphilitic, rheumatic, scorbutic, or malarious history should go to the Tropics, and persons either suffering from or having a tendency to heart disease should remain at home. With regard to age, no one should go to Africa under the age of twenty-five. Persons under this age are bound to suffer more from typhoid fever, from the severer forms of malarial fever, and from dysentery, than those who are older. With regard to women, apart from what has just been said, no woman with any tendency to the diseases special to her sex should go to the Tropics. With regard to phthisis, persons who have only incipient phthisis will do well in North and South Africa, and I am not at all sure that that disease should be a bar to their proceeding to Tropical Africa, provided that they are not going thither with the object of undergoing great physical exertion.

In going to Africa for the first time, it is well to arrive in the country in the coolest season. That season, as has been seen, varies considerably in different parts of the continent.

It is not my intention to refer to the diseases which, though occurring in Africa, are common to countries in the temperate zone. The outstanding diseases to which attention must be directed are yellow fever, malaria, dysentery, diarrhœa, and typhoid fever. With the exception of yellow fever, all these may be said to be prevalent throughout the continent. I will proceed to deal with the less important diseases first.

Ainhum.

This is a peculiar disease affecting the Negro race of both sexes. It always occurs in the small toes of the feet, and is diagnosed from leprosy and elephantiasis by there being no constitutional disturbance. The origin of the disease is unknown; it commences by an almost semicircular furrow in the digito-plantar fold on the internal and inferior surface of the root of the little toe. There is no marked inflammation or ulceration. The toe, however, increases in bulk until it is four or five times its ordinary size. The furrow gradually deepens until the toe hangs by a small pedicle to the foot. Sensibility of the toe is not lost. The treatment of the disease consists in amputation, after which it is not found that the other toes become affected.

Beri-Beri.

Beri-beri is a disease which manifests itself in anæmic or debilitated individuals; also in those following sedentary occupations. It is never met with in people until they have resided in the country, where it is endemic, for about a year. It is a fatal disease, death taking place either by syncope or from embolism. Its cause is unknown; there are, however, two theories in regard to it. The one is, that it is due to water impregnated by saline material, and the other that it is akin to Egyptian chlorosis (*q.v.*), and that water conveys to the individual the larvæ of the *Anchylostomum duodenale*. I incline to the former theory.

It has long been thought that a specific micro-organism must cause beri-beri, and two observers—Drs Mosso and Morelli—have examined the blood of eleven patients suffering from the disease, with uniform results. When rabbits and guinea-pigs were inoculated with cultures made from the micro-organisms of the blood, the animals apparently died from beri-beri, and after death three prominent conditions were found—ascites, hydroperi cardium, and nephritis. The liquid found in the abdomen and pericardium was strongly albuminous, and contained salts—corresponding, therefore, in general characteristics with the fluids found in the bodies of human beings suffering from beri-beri.

The symptoms are extreme weakness and prostration, dyspnœa and palpitation on exertion, numbness of the lower extremities, followed by œdema, anæsthesia, and, more rarely, by paralysis. The œdema gradually pervades the whole body, and effusions take place within the cranium, pleura, or pericardium. *Pari passu* with the increase of the dropsy, the whole body becomes numb. The urine is high-coloured and scanty, the specific gravity 1030 to 1040, and acid. It may be suppressed. Constipation is usually present. Considerable pain is often experienced in the cardiac region, and the pulse is irregular, unless effusion takes place within the cranium, when headache and delirium are frequent, and the pulse is slow and full. The prognosis is good if the disease occurs in young and otherwise healthy individuals, but the mortality on the whole is about 30 per cent.

The treatment consists in the use of iron, nux vomica and diuretics, and a stimulating nutritious diet. In India, Trecak farook is used with advantage. Frictions with rubefacient liniments are useful; petroleum, externally and internally, has been advised by S. Arokeum. The only preventive measure which can be recommended, apart from the ordinary precautions, is to ensure a good water-supply.

Bilharzia Hæmaturia.

This disease is caused by the ingestion of water containing the embryos of the *Bilharzia hæmatobia*, or possibly by bathing in water infested by them. It causes hæmaturia, cystitis, pyelitis, and sometimes dysentery. It is chiefly met with in Egypt and at the Cape of Good Hope.

It is quite possible that some minute leech-like animal fixes itself on the skin of the bather, and by means of an ovipositor implants the ova in some superficial cutaneous vein, and then the free embryos might be carried by the circulation from the ankle or leg to the pelvis. Persons who use river water, or water from pools or marshes, are most frequently attacked by the disease, those who use stored rain or well water being rarely affected. On examining the urine in these cases, it will be found to deposit a layer of dirtyish white flocculent matter containing short filaments of 135th of an inch in diameter, of a brownish colour and soft consistence. Microscopically, pus corpuscles are seen, and filamentous bodies containing great numbers of bright, highly refractive bodies imbedded in them. These bodies are the ova of the bilharzia. Stone in the bladder is not infrequently caused by the ova; the ova in the bladder become imbedded in a plug of hard mucus, and so form the nucleus of a stone.

Prophylactic Treatment.—After bathing in districts where this disease prevails, the skin should be thoroughly and very vigorously rubbed with a very rough towel. Filter and boil all drinking water. All raw salads and molluscous animals ought to be excluded from the diet. It is also essential to remove persons from the locality in which it is believed they have contracted the disease.

Treatment.—The patient must be well fed, must take moderate exercise, use cold baths and take tonics, either the mineral acids or the citrate of iron and quinine. Vesicle irritation must be subdued by the use of bicarbonate of potash and infusion of buchu. Hæmorrhage may be checked by uva ursi combined with small quantities of hyoscyamus, or hamamelis is very useful. Sometimes the injection of iodide of potassium, 3 to 5 grains to the ounce of water— and retention in the bladder for three hours—is beneficial. If the kidneys are affected, quassia, or the extract of male fern, should be administered. It is possible also that the following prescription may be given three times a day with advantage,—

℞. Bicarbonate of soda, 15 grs.
Chian turpentine, 10 grs.
Acacia mixture, 2 drs.
Chloroform water to 1 oz.

to which opium may be added if there is much pain or irritation in the urinary passages.

Delhi Boil, or Oriental Sore, or Biskra Button, or African Date-Mark.

This disease begins by itching, usually of some surface on the exposed part of the body. A papule is then formed, which soon becomes pustular. The discharge from the pustule forms crusts, under which ulceration goes on. The sore lasts for about five or six months, leaving deep dark cicatrices, hence the name date-mark. It is inoculable, and occurs at all ages. It is not accompanied by pain or fever. It is usually single, but may appear in crops. Although it is certainly a specific affection, it is as yet impossible to determine its cause. It does not appear to be due to water, or to climate, or to eating fresh dates. Some believe (Carter) it is due to a mycelium arranged in open and angular meshes with conidia on its free ends, having subsequently bright orange-tinted particles arranged in spherical or ovoid groups, supposed to be a further stage of development. It is met with in horses and dogs.

Preventive Treatment.—The use of pure filtered water, good food, absolute cleanliness of house, clothes, and person, avoidance of overcrowding, and contact with the disease in animals or human beings, with careful attention to sanitary surroundings, are the only means which can be suggested as prophylactic to the disease. The local treatment should be water-dressings, followed by linseed-meal or bread poultices; but when the ulcer is formed, a stringent lotion, such as sulphate of iron, carbolic acid, and iodine may be applied. If the patient suffer from malaria, quinine must be given. If a scorbutic taint is suspected, fresh vegetables and lime juice, or fresh lemon juice are advisable; or should a syphilitic taint be made out, small doses of mercury and iodide of potassium will be of use. In all cases a tonic regimen should be followed. Change of air and of drinking water is most desirable, and a voyage home may be needful in obstinate cases.

Dengue.

Dengue may be described as a specific fever, characterised by a high temperature, a peculiar and distinctive rash, violent and acute pains in the head and eyes, the muscles and joints; there is swelling of the joints, and the pains are apt to shift suddenly from one limb or joint to another; the throat and mouth are often affected, as well as the sub-maxillary glands; the sensorium is often much disturbed, and active, violent delirium is not uncommon. The disease is very infectious, and attacks persons of every age and sex. It may remit, and is liable to relapse. Some authorities consider that dengue is intimately connected with relapsing fever. Dr Christie thought that it was related to cholera. Although the complications of dengue are very troublesome, yet the prognosis is favourable. Deaths usually occur from syncope in children, in the aged, and in the debilitated. Summer and early autumn are the seasons when dengue is most prevalent. All epidemics have occurred during the hottest weather. It does not appear that the amount of moisture in the air has anything to do with its production. The onset of the disease is sudden. After suffering from severe rigors, the patient's temperature runs up to 103° or 104° F. Headache, pains in the joints, and nausea are prominent symptoms. The patient is sleepless and restless. Enlargement of the lymphatic glands is noticed. There may be epistaxis, salivation, diarrhœa, or dysentery. Hepatic derangements are sometimes seen. Women may suffer from uterine hæmorrhage, and if pregnant are liable to miscarriage. The first eruption, which appears on the third day, resembles that of scarlet fever, the throat being likewise affected. After two or three days the fever subsides, and the rash disappears, but after an interval of from 24 to 36 hours the fever returns, accompanied by a secondary rash resembling measles.

With regard to prophylactic treatment, no drug is of any service, but overcrowding and faulty sanitation appear to favour the spread of the disease. The greatest care should be taken in thoroughly disinfecting all fomites, for contagion may persist a long time, and the dead should be carefully disposed of.

General Treatment.—Castor-oil, or some warm carminative aperient, should be given at the onset of the disease, violent purgatives being contra-indicated. Then the following effervescing draught is useful,—

> R. Bicarbonate of potash, 2 drs.
> Tinct. opium, ½ dr.
> Syrup of orange, 6 drs.
> Camphor water to 6 ozs.

Half an ounce, with a tablespoonful of lime juice administered whilst effervescing, every three or four hours.

For the restlessness, 20 or 30 drops of chloric ether may be given with each dose, or saline, such as acetate of ammonia, or citrate of potash, with nitrous ether, may be useful during the pyrexia. If the temperature rises above 105° F., the patient should be sponged with cold water. If the pain is very severe, 10 or 15 drops of the tincture of belladonna should be given every three or four hours. Opium or chloral, or Dover's powder in moderate doses, are sometimes useful at bedtime. The eruption should be treated with simple bismuth ointment. Warm baths, in which a couple of pounds of bicarbonate of soda have been dissolved, are also beneficial. Liniments containing opium, belladonna, and chloroform may be applied to the spine and joints. After the first remission the following prescription should be given,—

> R. Carbonate of potash, 1 dr.
> Nitrate of potash, 1 dr.
> Tincture of orange, 3 drs.
> Water to 6 ozs.

1 oz. every three hours, and quinine 5 to 8 grs. should be administered thrice daily. For convulsions in children, bromide of potassium may be given. In adults, if there be great depression or marked nervous symptoms during convalescence, dilute phosphoric acid with nux vomica, or small doses of strychnine, should be employed. The enlarged lymphatic glands may give trouble during convalescence; they should be painted with strong iodine paint, and small doses of iodide of potassium should be given internally. The rheumatic pains are apt to persist, and it may be necessary

to blister the surface, and dress the blister with morphia; and if the internal administration of belladonna does not succeed in controlling the pain, the following prescription should be tried,—

> ℞. Nitrate of potash, 40 grs.
> Spirits of sweet nitre, 2 drs.
> Colchicum wine, 2 drs.
> Water to 8 ozs.

1 oz. thrice daily.

Guinea-Worm.
(*Filaria Medinensis*; *Dracunculus Medinensis*.)

The guinea-worm is a nematoid worm, the female only of which is known. It is about 1 mm. thick, and has an average of 2 feet in length; it may, however, be much longer. It is cylindrical; the anterior extremity is rounded, and presents a small depression surrounded by an elliptical chitinous plate, at the margin of which are two papillæ. The posterior end is a short curved point; it is opaque, of a milk-white colour, and on each side there is a longitudinal line. The interior of the worm contains an immense number of young filariæ in an acrid secretion. The tenacity of the tissue of the guinea-worm is considerable, so that a loop of the parasite will support the weight of 10 oz.

The largest number of persons becomes infested with this worm at the end of the rainy season, or in the hot season afterwards. The guinea-worm attacks all races and nationalities, both sexes, and at all periods of life. Usually only one or two worms are found in the patient, but I once saw one with no less than forty-two worms. It is said to gain an entrance into the body by means of drinking water, but this is doubtful. So far as I know, there is no case on record in which the worm has been found within the abdomen. It is in all probability deposited beneath the skin, and this theory is supported by the fact that it only occurs in those who walk bare-footed either constantly or occasionally, that it principally occurs in the lower extremities, that in those cases in which it occurs on the trunk or arms it is in persons who have slept on the ground, or who have carried earth or water. In 1190 cases I saw, the feet were affected in 556 cases, the legs in 274, the thighs in 104, the scrotum in 4, the penis in 3, labia majora in 3, abdominal walls in 1, the breast in 2, the back in 143. I never myself saw it in the head, arms, or elsewhere. The average time of incubation after the ova is deposited within the skin appears to be from three to six months. It is only when the worm arrives at maturity that the patient first becomes really aware that he has contracted the disease. He suffers from a severe attack of fever, the stomach becomes irritable, bilious vomiting takes place, and attention is now first attracted to the situation of the guineaworm. Intense itching is felt, with a sensation of a thin cord underneath the skin; occasionally also a small pimple or blister can be seen, and when this occurs in a district infested by the guinea-worm, it may generally be regarded as a sufficiently diagnostic sign, especially if accompanied by any swelling.

The development of this characteristic blister or vesicle always coincides with the advance of the worm to the surface of the body. The blister may sometimes be as large as half a walnut, and may be attended also by an eruption resembling nettlerash. When this vesicle is opened, it is seen to be filled with either a glary, whitish fluid, or with the reticular portions of the true skin, the areola being a meshwork filled with serum, at the centre of which a small aperture is visible, in which the extremity of the worm will be found. As the worm protrudes from the skin, it should be secured to a small piece of twig or crowquill and gradually withdrawn. Great care must be taken not to break the worm, as if it is broken extensive and destructive inflammation in the connective tissue occurs. The prognosis in general is good if the case be treated carefully, but it may cause distortions of the lower extremity, such as talipes equinus, permanent enlargement of the internal maleolus, permanent contraction of the leg or the thigh, and sometimes permanent anchylosis of the knee joint, or gangrene of either toe, foot, or leg.

Preventive Treatment.—Never walk bare-footed in a region where the worm is endemic; do not bathe in muddy pools.

After walking through swamps, see that the skin is well rubbed with a rough towel or flesh brush. All drinking water must be filtered.

Internal Treatment.—Nitrate of potash in 2 dr. doses given in butter-milk is said to cure guinea-worm in from three to five days, and the application of electricity to the worm-affected part may cause its death. Asafœtida in 5 to 15 gr. doses daily for a week has cured many cases. It is said also that feeding a patient on sugar-candy for twenty-four hours, without any other food or drink, may cause the death of the worm.

Extraction of the Worm.— After securing the worm, attempts should be made to extract it every twenty-four hours. As much as 6 or 8 inches a day may be extracted and wound upon the quill, which should then be fastened parallel to the limb with two pieces of strapping, and the part dressed with lint soaked in a solution of alum, 8 grs. to the ounce. This prevents the part becoming dry, and also strengthens the worm, and so tends to diminish the danger of its breaking. The natives are very skilful in extracting the worm by making an incision over it and rapidly turning it out, but this requires great skill.

Constitutional symptoms, when they arise, must be treated on general principles, and if malaria complicates matters, the use of quinine must be energetically pushed.

Egyptian Chlorosis.

This disease is common in Egypt. It is caused by the ingestion of the *Anchylostomum duodenale,* from which it is reported that a fourth part of the population suffers. It is said to be a stage in the development of the *Dochmius trigonocephalus* of the dog. It attaches itself to the lower portion of the human duodenum and to the jejunum. The symptoms caused by the presence of this worm are those of pernicious anæmia. For debilitated individuals it may prove fatal in a few weeks. In well-fed persons the disease may exist for two or three years. It is found that the best treatment to get rid of the worms is the administration of the milky juice of the

Ficus doliaria or of the *Carica dodekaphylla*, or Thymol is often of great use. Dr Sandwith gives to adults 2 grms. at 8 A.M. with 25 grms. of brandy, and repeats the dose at 10 A.M. At noon a dose of castor-oil should be given. Care should be taken in very debilitated subjects. If necessary, this treatment may be repeated in a week. It is well to remember that the patient should keep perfectly still after taking the medicine, as giddiness and faintness are apt to ensue. The brandy is given both to dissolve the drug and also to prevent collapse. The subsequent treatment of the anæmia which has been produced must be carried out upon general principles. It will be found that Levico water is very useful in improving the condition of the blood, but as it would not be easily carried, iron and arsenic should be given thrice daily.

Snake-Bites.

One or two hints with regard to the prevention of snake-bites may not be out of place. In order to prevent snakes from entering a house, it is advisable to have a path four or five feet wide encircling it, and covered with rough stones. Keep the verandahs free from frogs, especially during the wet season. A frog is a temptation which a snake has little or no power to resist (Waring). Place a coil of camel's-hair rope round the bed; snakes will not cross it. Never get out of bed during the night with bare feet without a light and first seeing if the way is clear. If a snake is seen coiled up or in an apparently lifeless state in the road, it should be avoided, as it is probably only torpid with cold, not dead (Dr Chevers). It is well to remember that a poisonous snake-bite may be diagnosed by the *two* well-marked wounds made by the fangs. In treating snake-bites, it is probable that the hypodermic injection of strychnine, as recommended by Dr A. Müller, is the best treatment we possess. "Nothing less than 16 ms. of the liquor. strychniæ (B.P.), in very urgent cases even 20 or 25 ms., should be injected into any person over fifteen years of age. Even children may require these large doses, as they are determined by the quantity of the poison they have to counteract, and are kept in check by it.

The action of the antidote is so prompt and decisive that not more than fifteen or twenty minutes need elapse after the first injection before further measures can be decided on. If the poisoning symptoms show no abatement by that time, a second injection of the same strength should be made promptly, and, unless it is followed by a decided improvement, a third one again after the same interval. As the action of strychnine, when applied as an antidote, is not cumulative, no fear need be entertained of violent effects suddenly breaking out after these large doses repeated at short intervals."

Yaws or Framboesia.

This disease consists of an eruption of yellow or reddish-yellow tubercles, which gradually develop into a moist exuding fungus, without constitutional symptoms, or with such only as result from ulceration and prolonged discharge, namely, debility and prostration. Its predisposing causes are filth, vitiated atmosphere, and want of animal food in a tropical climate. It is more common among the coloured than the white population. It is epidemic and contagious by actual contact. The period of incubation ranges from three to ten weeks. It is not liable to recur. Its duration is from two to four months, but it may last for a year. It frequently runs in families, and is apt to be communicated by clothing, especially by boots. Children are most subject to it, then men, lastly women. The disease often begins with a severe febrile attack; in a few days small spots appear, principally on the face, in the axilla, in the neighbourhood of the groin, or on the feet. They increase gradually until they are as large as a pin's head, the surrounding skin acquiring an unhealthy aspect. In about a week these little tubercular swellings exude a thin sanious fluid, forming dry scales or scabs. The surface remains covered with these scabs for a week or ten days, if undisturbed, during which time a fungoid excrescence grows underneath it, so as to form a projecting mass one or two inches in diameter. The skin around is hard and firm. Crops of yaws arise at different periods. After maturity they may remain stationary for some weeks. One excrescence in each group is generally larger than the rest, and is called the "mother" or head yaw. The disease runs a definite course, exactly like the exanthematous eruptions. When the ulcerations heal they leave a pigmented stain, but the "mother" yaw leaves a large scar.

Treatment.—The preventive treatment is cleanliness, good diet, and the avoidance of contact with persons suffering from the disease or with their clothing.

The general treatment must be guided by the symptoms present, as the disease cannot be abbreviated. Full animal diet should be given, and perfect cleanliness enjoined, with exercise and plenty of fresh air. Tonics and alteratives are required from the first; arsenic is extremely useful, so are the mineral acids, sarsaparilla, and bark. Iodide of potassium also is given, in combination with liquor arsenicalis and alkalis, and is exceedingly useful when the ulcers are indisposed to heal.

With regard to local applications, carbolic acid solutions, or dilute nitrate of mercury ointment, or creosote in the strength of 1 dr. to 1 oz. of lanoline, should be employed. Authorities differ as to the administration of mercury in this disease; it should at any rate be avoided in debilitated subjects.

Elephantiasis Arabum.

This is a chronic disease, which may be said to be characterised by an enormous hypertrophy of the skin and subcutaneous tissue, caused by recurrent inflammation of the vessels and lymphatics in the part affected. It is unnecessary to refer at any great length to this disease, because it is very rarely that white residents in Africa are affected by it. Various parts of the body are attacked—the legs, scrotum, pudendum, abdomen, and breasts; most chiefly, however, it is found affecting either the legs or the scrotum. Males are most frequently attacked at about the age of puberty. It is non-contagious; it is not hereditary; its cause is unknown.

The treatment, when the disease has once become manifest, is removal from the area in which it was contracted. It was once supposed that tying the femoral artery would cure the growth in the leg, but that treatment is unsatisfactory. Persistent strapping from the foot upwards has also been recommended, but it likewise does little good. The scrotum may be removed, and even enormous tumours weighing 40 to 60 lbs. are often successfully treated in this manner. In a case I treated lately, in which the disease was limited to the body and the thighs as far as the knees, I obtained a cure, by enjoining absolute rest, giving the patient a hot bath every day, a moderate amount of food, chiefly milk, having the patient regularly massaged, and by applying the constant current for twenty minutes each day. A mixture was prescribed containing quinine, arsenic, iron, and strychnine, and the bowels were regulated by the frequent administration of aperients.

Leprosy.

This is a disease caused by a bacillus which is chiefly found in the exudation cells, but also in the diseased connective tissue, more rarely in the blood-vessels. There are three varieties—tuberculated, non-tuberculated, and anæsthetic. The recent Commission which has been held on the subject does not believe that the disease is either contagious or hereditary. It is practically incurable, but benefit may be obtained by the internal and external administration of chaulmoogra oil, and quite recently extract of the thyroid gland has been given with marked success. It has been stated that the production of leprosy is due to extremes, frequent and rapid transitions of temperature, but it is not so. Various articles of diet have been blamed for its cause —fish diet, salt or rotten fish, immoderate use of pork, and the use of decomposing rice or maize, but none of these articles of diet can be its exciting cause.

In Central Africa the natives certainly believe that the disease is contagious, and they also believe that sleeping in a hut which has been inhabited by a leprous patient is dangerous. It is, indeed, necessary to avoid contact with lepers as much as possible. White men in Africa rarely, if ever, suffer from this disease.

Yellow Fever.

Yellow fever is a pestilential contagious disorder of a continuous and special type, depending for its origin and spread on a temperature not lower than 70° F. As a general rule it occurs but once in a lifetime. Its spread is favoured by the gathering together of persons born in a cold climate.

Etiology.—Yellow fever is entirely distinct from malaria. Its production requires a temperature of from 68° to 70° F. When once originated, however, an epidemic may spread at a lower temperature, but it dies out if the temperature falls to freezing point. The influence of moisture in the air constitutes a second factor in the production of yellow fever. Abundant continuous rain does not infrequently bring an epidemic to an end, probably by modifying the temperature, but a certain saturation of the atmosphere is an essential condition for the production of the disease—probably 74 per cent. of moisture. Epidemics cease when the amount of moisture is as low as 58 per cent. The only influence which wind has on yellow fever is by its modifying the temperature. The disease rarely leaves the sea coasts and the shores of large rivers; it arises in the filthy quarters of towns, in the centres of poverty where the people are densely crowded. The geological characters of the soil have apparently no connection with the production of the disease. It is most interesting to notice the influence which circumstances of race, nationality, and acclimatisation exert upon the disease. Where it is endemic or epidemic, newly-arrived strangers, or such persons as have not yet become fully acclimatised, are the persons who suffer most. This is well seen if a large body of troops or a shipload of emigrants arrive at any place where the disease already exists, though it may be very mildly. An epidemic at once springs up, and the new arrivals are the first persons attacked. The degree to which this proclivity of strangers exists will depend to a great extent on their nationality, that is, on the mean annual temperature of their native country. The liability to attack, as well as the mortality amongst the newcomers, bears a close relation to the distance from the equator of their place of birth. Although no absolute immunity is acquired by acclimatisation, yet it is true that a certain amount of immunity is possessed by those who have lived for a considerable period in any locality constantly or frequently visited by the disease. The chances of immunity appear to be always in direct proportion to the length of residence at the headquarters of the disease, but no protection is acquired except by those who have passed through a previous epidemic period without quitting the country. Any benefit, however, gained by acclimatisation is immediately lost on change of residence, even though that change be to a healthier locality. This remark applies equally to negroes and white men, but negroes suffer far less from yellow fever than do the whites.

The nature of the yellow-fever poison has been found by Dr Domingos Freire to be a specific cryptococcus. He has also found out that this micro-organism secretes an alkaloid resembling a ptomaine, which acts as a violent poison. With regard to contagion, there is no doubt that it may be conveyed by fomites or merchandise, as also by ships, and it can be transported farther by sea than on land. Hence the necessity during an epidemic of completely isolating vessels in a harbour. It does not appear that contact with the sick has power to spread the disease. Electricity appears to have a singular and baneful influence upon persons suffering from yellow fever.

It is impossible in the space at my disposal to detail the symptoms of yellow fever, but there are a few well-marked symptoms which deserve notice. First, the attack is sudden, there is a want of correlation between the pulse and temperature, albuminuria is invariably present, the patient suffers from the black vomit, from a general hæmorrhagic tendency, and from a yellow discoloration of the skin, often, too, from suppression of urine.

Prophylactic Treatment.—Avoid the yellow-fever season if possible, namely, the hot and rainy season; avoid chills by wearing proper clothing; avoid exposure to the sun as much as possible. Individuals attacked by the disease should be isolated. Residences should be chosen at the highest altitudes possible, in any case the second storey is preferable to the first, and in camp the ground should be disturbed as little as possible. Strictest disinfection should be employed. The water and food supply should be well cared for, and it should be remembered that the yellow-fever bacillus has been found in the soil. Exposure during the night is inadvisable, for Carlos Finlay has demonstrated that the disease can be communicated by the mosquito. Yellow fever is also a disease in which quarantine is necessary and effectual; not only is it necessary to place persons coming from an infected place under observation, but their clothing and goods should be thoroughly disinfected.

Inoculation.—Since the researches of Freire and Finlay, it is possible to employ protective inoculation against yellow fever. Finlay allows a mosquito to bite a yellow-fever patient and then a healthy person; a mild attack of yellow fever is induced, protecting the person thus treated from a subsequent attack. Dr Freire, having isolated the yellow fever micro-organism by a series of cultivations, attenuates it and produces a fluid which almost entirely protects persons from yellow fever. In no case has the inoculation been harmful, and the mortality of those inoculated was only rather more than 0·4 per cent. in 10,881 cases inoculated (1890), showing that this procedure confers almost certain immunity from the disease.

Treatment.—With regard to the treatment of yellow fever, little can be said. It is important that each patient should be allowed at least 2000 cubic feet of space; the room should be kept at an equable temperature, and the patient protected from draughts; indeed, many advise the treating the patients in tents or in the open air. Absolute rest in a recumbent position must be rigidly maintained. The patients must be lightly but warmly clothed, heavy blankets being avoided. Doctors and nurses should be cheerful, and encourage the patients as much as possible.

With regard to drugs, no specific for yellow fever is known, and, practically, symptoms must be treated. I should be inclined myself to recommend either of the two following treatments. The first is recommended by Nelson. He gives

15 grs. of quinine, half an ounce of sulphate of sodium, dilute sulphuric acid, and tincture of cardamoms, at first. If after the first two days the temperature remains above 100° F., with the usual symptoms of yellow fever, he adds phosphoric acid largely diluted with water, every hour or two. Diaphoresis is induced by vapour baths; the diet consists of iced milk and beef broth in small quantities at frequent intervals. On the other hand, Sternberg recommends bicarbonate of soda 150 grs., bichloride of mercury ⅓rd gr., water 2 pints; an ounce and three-quarters to be given ice-cold every hour. This treatment is slightly modified by Mitchell, who increased the dose of the bicarbonate of soda to 4 drs., and the bichloride to half a grain. When patients are thus treated from the first day, vomiting rarely occurs. Diuresis is well maintained. After the eighth or tenth day it is necessary to suspend the bicarbonate of soda and give stimulants, and to combat the adynamia and the hæmorrhages, etc., with the customary measures. For the vomiting I believe turpentine is the best remedy; it may be administered either by the mouth or by enemata, and the body may be rubbed with a mixture of turpentine and olive oil. If suppression of urine occurs, I know of no better treatment than to apply a digitalis leaf poultice to the loins, and to inject into the rectum a pint or more of ice-cold water at regular intervals.

The discharges of the patient are best disinfected with either chloride of lime or perchloride of mercury. Bedding and clothing are better destroyed by fire. Hospital wards or the holds of a ship should be fumigated with nitrous acid for at least forty-eight hours, and then all the woodwork washed with chloride of lime. For disinfecting the bilge-water of ships, chloride of lime must be employed, or, better still, the bilge-water should be pumped out.

Typhoid Fever.

I have only a few remarks to make on this subject. Typhoid fever certainly exists in Africa, and it is also certain that the death-rate is higher there than it is in more temperate zones. I hold the view that the disease is due to

Eberth's bacillus, and I have only a few words to say with regard to the prevention of the disease.

Typhoid fever is most prevalent during the hottest months in Africa, and it should be remembered that sandy soil favours its spread, as the dried excreta of patients may be conveyed by the wind unless care be taken. The utmost care should be taken to ensure the fullest sanitary precautions. The excreta must be properly disinfected, and the water-supply should not only be carefully selected, but all water should be filtered and boiled. Milk, too, deserves special attention. The meat-supply should also be investigated, and all tainted supplies rigorously rejected. It is necessary also to pay attention to the vegetables, as undoubtedly they may carry the infection. All patients should be thoroughly isolated, and their bedding and linen destroyed. It is very necessary, in my opinion, to get rid of the idea that typho-malarial fever exists, and in cases of doubt a bacteriological investigation should be made, which failing, the disease should be treated by quinine, and if it does not succeed in reducing the temperature, then the case should be treated as one of typhoid fever. The cases which have been designated typho-malarial fever are in reality severe cases of remittent fever lapsing into a typhoid state, or else enteric fever modified by its occurrence in a patient who has previously suffered much from malaria, or occurring simultaneously with an attack of malarial fever (Duncan).

Tropical Dysentery and Diarrhœa.

Dysentery has practically the same distribution as malaria in Africa, and there are only some minor differences met with in the distribution of the two diseases. It does not always follow that the maximum intensity of the diseases coincides.

In referring to the etiology and prevention of dysentery, I may, to economise space, include diarrhœa as well, for, although I believe true tropical dysentery to be due to the amœba discovered by Cartulis of Alexandria, which discovery has been confirmed by American observers (see

Johns-Hopkins Hospital Reports), yet both diseases may to a certain extent be combined, and the precautions necessary to avoid the one are those which would prevent the other.

Both diseases are most prevalent in the hot and rainy seasons; both are liable to be produced by rapid alternations of temperature and by chill. Therefore persons in Tropical Africa should avoid chill by means of careful clothing, and by the invariable use of a cholera belt. Excessive exertion also predisposes to both diseases, and both are especially met with in damp, swampy places, and in all districts where the soil is impregnated with decaying vegetable débris. The drinking water should be as pure as possible, and in cases where the water-supply is doubtful, it should be filtered and boiled. All stagnant water should be, if possible, avoided. It is also of importance to remember that both a monotonous diet and salt rations frequently induce diarrhœa, and predispose to dysentery. Unripe fruit, and especially over-ripe fruit, should be avoided. There is no doubt that in Africa many cases of diarrhœa and dysentery are induced by exposure to the night air, and also by sleeping on the ground. Where they are prevalent, it is well to isolate the patients, and to carefully disinfect their excreta; and finally, it must be borne in mind that malaria may complicate both diseases, and that then, unless the malarial factor is taken into account, the disease cannot be cured.

One may summarise the predisposing causes of dysentery and diarrhœa as follows:—Frequent exposure to malaria, great bodily fatigue or excessive anxiety and mental distress, excess in the use of alcohol and tobacco and narcotics, overcrowding, the use of tainted food or the prolonged employment of salt provisions, and lastly, the too frequent employment of strong purgative medicines. The exciting causes of these diseases are—unwholesome drinking water, the use of indifferent food, great and sudden vicissitudes of temperature and chill, impure air, intestinal worms, and abscess of the liver.

Nothing need be said as to the treatment of diarrhœa, as this must be carried out on general principles; but with regard to dysentery, my experience points to the advisability of treating it in Africa with large doses of ipecacuanha. This I consider the only treatment of any practical value. After sending the patient to bed, a mustard poultice should be applied to the epigastrium, and 30 drops of laudanum given at once. After half an hour 30 or 40 grs. of powdered ipecacuanha should be given, in as small a quantity of fluid as possible. A similar dose may be repeated in twelve or twenty-four hours if necessary; after this, during the succeeding days, the dose should be gradually lessened to 10 or 15 grs. a day, until the patient has perfectly recovered. In very severe attacks, as much as 2 drs. of the powder have been given without producing vomiting. Fairly large doses of quinine are required in all cases of malarial dysentery. In the treatment of the scorbutic form of dysentery, lime juice, fruit, and vegetables should be given, with as much animal food as the stomach will bear. In treating natives in Africa, the great difficulty is to ensure proper diet, for, unless the patient is carefully watched during convalescence, a relapse will follow the least indiscretion. A sea voyage is beneficial when a patient is convalescent, but it is not to be recommended during the continuance of the attack.

Malaria.

There are few regions in Africa where malaria is not a scourge, and those few have been indicated in my survey of the various African regions.

I may say at the outset that I believe malaria to be produced by the hæmatozoon discovered by Laveran. His researches have been confirmed by observers in Europe, India, America, and Africa. The life-history of the hæmatozoon we do not know, and therefore we can only state that it requires a mean summer isobar of 58°–60° F., and considerable moisture; also, other things being equal, the greater amount of organic matter in the soil, the more virulent will the production of the disease be.

Of the various types of malarial fever, the intermittent is the most widely distributed, the remittent and pernicious fevers only being met with in comparatively limited areas, and in Africa these are found upon the coasts, along the rivers, and in the water-logged swampy districts at an altitude of under 3000 feet. The quotidian and tertian types of intermittent fever are the ones most frequently met with. The type of fever stands in a definite relation to the intensity of the malarial process; thus we find that the tertian type prevails in those regions of Tropical Africa where the malarial process, although indigenous, is more sparingly produced. The frequency of the occurrence of the quotidian type of fever, either in endemic areas or in epidemics, is in direct proportion to the severity of the process. When an epidemic wave of malarial fever passes over a district, the tertian type is seen at its outbreak, whereas at the height of an epidemic, or whenever it assumes a severe character, the quotidian type obtains; and as the outbreak of sickness abates, one meets with a return to the types of fever having a longer interval between the paroxysms. In the higher latitudes in Africa, and also in the higher altitudes, the quartan type of fever makes its appearance.

All races suffer from malaria, although the Negroes suffer less from it, always provided that they do not migrate. In Africa, as in all parts of the world, strangers suffer more severely from it than does the indigenous population. The incidence of malaria is, to a certain extent, governed by the seasons. In those places where it is endemic, it occurs all the year round, but where it is only slightly developed there are two maxima, one in spring and one in autumn, and a considerable decrease in the disease in the interval. In Africa, in the worst malarious regions, the disease is practically most rife at the beginning and at the end of the rains. The relation which malaria bears to heat is as follows: the greater the mean summer temperature (moisture, etc., of course being taken into account) the more malaria, the amount of malaria decreasing with the mean annual temperature of the place.

The influence of rain or moisture has undoubtedly much to do with the production and spread of malaria. With reference to the rains, the malarial poison is most virulent either when they set in after a long period of heat, or when the rains cease and give place to warm dry weather. An endemic outbreak of malaria and its epidemic spread are both notably diminished at the height of the rains, if they are very abundant, but the malarial process is developed more abundantly in wet than in dry years. These remarks are well illustrated by the behaviour of malaria in different districts. In Equatorial Central Africa, where the rainfall is fairly equally distributed throughout the year, the amount of the disease remains practically the same, but in regions, *e.g.*, along the White Nile to the north of Lado, where there are two wet seasons, a rise and fall in the production of malaria is manifest. But it is not alone rainfall which influences the production of the disease. Drainage from rivers, lakes, and pools, periodical or irregular inundations, and the height of the sub-soil water, influence its production. This last point is of importance, because it explains the occurrence of malaria in localities remote from river basins, in the Sahara, in Darfur, etc.

Although the geological characters of the country would appear to exert little or no influence on the production of the disease, it is the contrary with the physical characteristics of the soil. Clay, loam, clayey marl, and marshy soil are most favourable to the production of the disease. A porous chalky soil is less favourable, and a sandy soil least so, provided that they do not rest either upon clay or firm rock. Again, the greatest amount of malaria will be found where the organic matter in the soil is greatest. It is also an undoubted fact that changes in the soil, produced by cultivation or its neglect, influence the production of the disease. In well-cultivated countries malaria disappears, and if marshy districts are well drained or completely covered with water, the disease is also diminished.

The configuration of the ground also causes a local effect, for it is found that the disease is more virulent in the lowest altitudes; even the difference of 50 or 100 feet in altitude in a plain makes a considerable difference as to the salubrity or otherwise of a given spot.

Winds act only indirectly on malaria, as, for instance, by moderating temperature; they may, however, act directly in the diffusion of the poison or in preventing it exercising its potent effects. Wind may carry the malarial poison from a marsh to a distance of some two or three miles. Malaria may rise to a height of 600 or 700 feet in a calm atmosphere; wind will prevent this vertical diffusion.

Water can convey the malarial poison, but it is unknown at present how far it can carry it.

The poison is ponderable, and affected by barometrical pressure, and it is possible also that food may be contaminated by it.

The influence of jungle and forest on malaria must also be noticed, because so much of Central Africa is covered by one or other. In a jungle, malaria is intensely virulent, and, owing to want of ventilation by the penetration of winds, it is there in a very concentrated form. In forests the production of malaria is to some extent lessened by the shade, and by the trees diminishing the amount of rainfall reaching the soil. There is no doubt that forests often act as a screen or filter, and therefore protect the district from malaria when they lie between it and a marsh.

With regard to the prevention of malaria, much may be done by careful drainage, not only of the surface, but of the sub-soil water. Great care should be exercised in the choice of a residence, ravines being avoided, also the neighbourhood of swamps. Settlements, and even individual houses, should be on the most elevated situations, and it should be remembered that malaria is less rife in the centre of towns, especially if the streets are narrow and crooked. The proposal to build houses in the form of a hollow square is to be commended, and in all cases they should be constructed with a blank wall to the prevailing wind, especially if that wind blows over a marsh. The thick jungle in the neighbourhood of a settlement should be destroyed, but care should be taken not to remove either thickets or trees between a settlement and a marsh. The ground under and around a habitation should be rendered impervious to water and air, and the sleeping rooms should be in the second storey. In camping out even, considerable protection may be obtained by sleeping in a mosquito curtain in a hammock slung between two trees; this is far preferable to sleeping on the ground. A good deal may be done to make a settlement healthy by planting large trees, the eucalyptus, etc.; and Martin Clark recommends the plantation of bananas in the reclamation of malarious lands.

With regard to personal hygiene, food should be taken in sufficient quantity, and it is a mistake to think that white races in the Tropics can exist on native food. They should, however, not consume as much animal food as at home. Water must be boiled and filtered, and milk also boiled. Coffee apparently acts as a prophylactic to some extent. Moderate smoking is advisable. Strict temperance must be the rule, and persons must protect themselves as far as possible from chill, for although chill does not produce malaria, it may act as the exciting cause of an attack. The night air should be avoided, because then the malaria becomes concentrated, on account of the air cooling more rapidly than the earth.

With regard to the use of drugs, quinine is certainly to some extent a prophylactic, and should be taken in doses of 3 or 4 grs. daily during, and for fourteen days after, special exposure in malarious regions; but I do not think it advisable to take the drug continuously, for in my experience the system becomes habituated to its use, and, as it will not entirely prevent attacks of malaria, larger quantities are required to cut short the attacks when they occur. Another plan I have found successful is to give 15 grs. of sulphate of quinine twice a week for six weeks, and then 3 grs. daily for a month. Quinine should not be taken in either tea or coffee, and the drug should not be given in the form of pills. The use of lemon juice is very beneficial, and arsenic in minute doses may likewise be employed with advantage.

With regard to the treatment of malaria, I believe Laveran's recommendation to be the best. For the first three days administer 12 to 15 grs. of hydrochlorate of quinine daily; from the 4th to the 7th days omit the drug; on the 8th, 9th, and 10th days give 10 or 12 grs. daily; from the 11th to the 14th days omit the drug; on the 15th and 16th days give the same dose, and again on the 21st and

22nd days, omitting it on the 17th to the 20th day. In remittent fever the quinine should be given when the temperature falls, however small the fall may be. In pernicious fevers a hypodermic injection of the drug is indicated. I think the bisulphate of quinine, with a little carbolic acid, and glycerine and water, at a temperature of 100° F., is the best solution to use for this purpose. In severe remittent fever I consider Warburg's Tincture exceedingly useful, but I invariably prescribe it in the tabloid form.

Hæmoglobinuria, or Black-Water Fever.

This condition may occur as a complication in a malarial fever, or it may occur as a disease apparently unconnected with malaria. It may be due either to the disintegration of the red blood-corpuscles in the liver and spleen—the products being eliminated by the kidneys; or the kidneys may be congested, and a dissolution of the hæmatic elements in the kidneys results. The patient passes " porter-like urine," small in amount. Death is usually due to collapse following suppression of urine, and there may be convulsions.

Each case must be treated on its own merits. In some, quinine is indicated, dry cupping, and hot fomentations over the loins. Diuretics are indicated in other cases—acetate of potash with squills. As Dr Eyles points out, patients suffering from this complication are apt to be panic-stricken.

I cannot conclude without stating it to be my definite opinion, that as in India, so in Africa, the progress of medicine and hygiene will before long conquer most of the obstacles to the civilisation of that continent.

APPENDIX.

A New Method of Illustrating the Geographical Distribution of Diseases. A Paper read at the International Congress of Hygiene and Demography at Budapest, September 1894.

For some years I have felt the want—not alone for teaching, but also for many other purposes—of a method of illustrating, clearly and fully, on a map, the Geographical Distribution of Disease.

The methods in general use—cross etching, shading in colours, etc., were at the best unsatisfactory, were capable of only a limited application, or required many maps to illustrate a district fully.

By the use of symbols, as in the map which accompanies this work, I trust I have improved upon these previous methods. I have at least removed some of the disadvantages attending the adoption of them for illustrative purposes.

Thus, by the symbol method, not only can I show the Geographical Distribution of most diseases,—even all, if the map is large enough,—but I can illustrate graphically (by the repetition of the symbols) the degrees of severity of disease in any particular district.

Thus, we have the colour and shape of the symbol, to indicate the disease which is present, and the number (1, 2, 3) of symbols grouped together, to indicate the degree of the prevalence of that disease. For instance, the Geographical Distribution of Disease in Cape Colony is in our chart illustrated thus:—

> One blue dot—denoting the occurrence—though to no great extent— of Leprosy.
> Two black stars—denoting the prevalence of Syphilis.
> Three red stars—denoting that endemic Hæmaturia is not only prevalent, but is *very* prevalent.

Again, in a map constructed on this method, a view may be at once obtained as to the comparative salubrity of any area.

Thus, if one looks at the Zanzibar area, it is seen at a glance that no less than eleven diseases are indicated to be *very* prevalent; five to be prevalent; and only one—Measles —to be unimportant: we may therefore conclude that the East Coast of Africa is very unhealthy. Needless to say, the nature of the diseases present will influence our opinion, but enough has been said to illustrate the point.

It may be well to point out here, what is probably self-evident, that for Malaria, the chief disease of Africa, a tint has been employed, and the three densities of colour show the intensity of the malarial process, whilst its distribution is at once obvious.

We admit that many diseases are either caused by, or at any rate influenced by, climate, and therefore I have introduced upon the map sufficient climatology to illustrate this point. The mean annual temperature is represented by figures, thus, 80° F.; the mean annual range of temperature, thus (10° F.); the annual rainfall in inches, 50 inches; the annual relative humidity, 70 %; the annual range of humidity (15 %); altitude, 950 feet; and the prevailing winds, by symbols, as seen on the map.

A map, then, constructed upon the lines I have thus briefly laid down, will at once give us the answer to a large number of questions.

1. What is the climatology of an area?
2. What are the prevailing diseases?
3. Are these diseases due to climate or not?
4. Are they influenced by altitude or not?

Note.—This last question may perhaps be better answered by placing my map alongside one showing altitude, or by introducing contour lines upon it.

Then, if the area mapped is a tropical one:—

1. Can Europeans reside in a given area?
2. What precautions should be taken by travellers, proposing settlers, or armies proceeding there?

3. Is it likely, from the character of the diseases present, the climatology and the altitude, that a given area may be rendered habitable by Europeans?

These problems might be added to, but I have, I think, given enough to illustrate my view, and I must be brief.

I have chosen the map of Africa to illustrate my method because of the interest Africa always excites, and for this reason as well, that it has all varieties of climate.

I have not overloaded the map with detail, so that it may the more clearly illustrate my theory, and I do not profess that every disease is represented which may occur; still, broadly speaking, it is correct.

The map is divided into areas having, as nearly as may be, the same climatology, and the diseases of each area are grouped together.

I will now take two or three of the questions given above, and it will be seen how an answer can be found by a glance at the map.

Why is the geographical distribution of malaria in Africa such as that represented on the map?

Apart from Northern Africa, malaria is most severe, as is seen, along the East and West Tropical coast-lines, for there heat and moisture are met with in abundance, besides which the coast-lines are low and swampy.

Malaria decreases generally as we leave the coast and proceed inland. Why? Because we have an increased altitude.

There is no malaria in an area near Tete, nor in four areas, two on each side of the Victoria Lake. Why? Because the altitude—over 3000 feet—is too great for its production. (*Note.*—This is a general statement, local conditions may modify it.)

Malaria is generally absent from the Sahara, because, though low-lying, there is little or no moisture, and the heat is very great. (A few spots of malaria are seen; these represent oases where the disease may occur.)

Altitude, temperature, and probably also winds, render the Cape free from malaria.

Where might Europeans probably colonise in Tropical

Central Africa? The answer is almost self-evident—in those four white areas, where malaria is absent. These areas, again, could be extended by hygienic measures when we consider the configuration of the country.

Again, looking at the map, we find that with very few exceptions (and these could be explained by local conditions or importations of the disease) phthisis is absent where malaria is most prevalent. This is partly explained, no doubt, by altitude, but more probably, as I believe, by a certain antagonism between the two diseases.

This will, I think, be enough to show what can be read with ease from a map like this.

In detailed maps or smaller areas, of course much more may be done, and many more niceties of detail could be introduced.

| Malaria, |
| Typhoid Fev |
| Typhus Feve |
| Relapsing Fe |
| Cholera, |
| Dysentery, |
| Diarrhœa, |

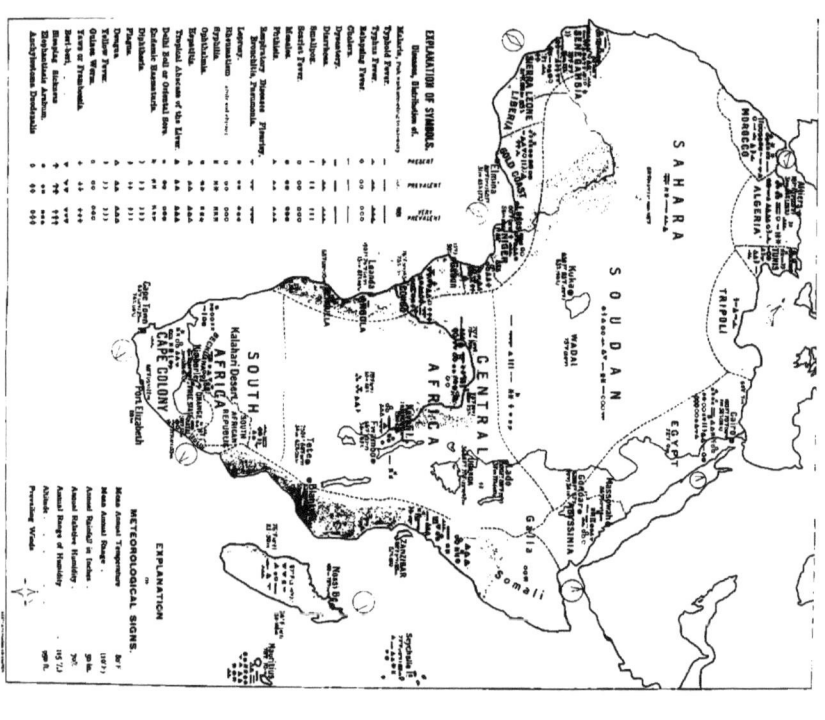

DISTRIBUTION

BY

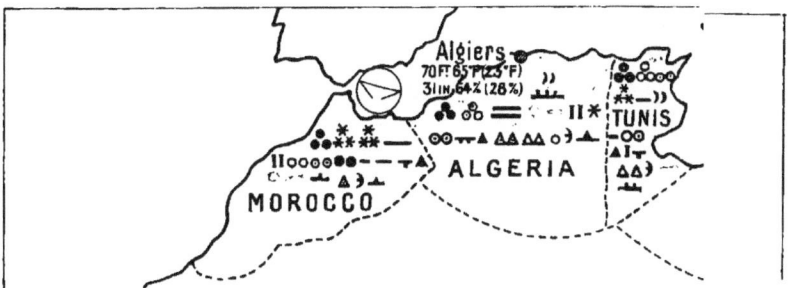

Distribution

BY

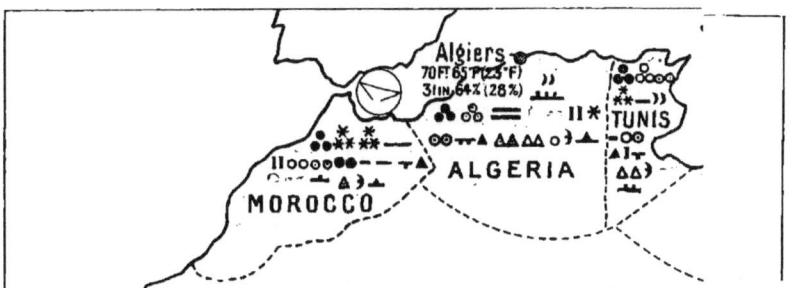

TABLE showing the Distribution of various Diseases in different Areas in Africa. The Amount of Disease is illustrated by the number of stars (*).

www.ingramcontent.com/pod-product-compliance
Lightning Source LLC
Chambersburg PA
CBHW031604110426
42742CB00037B/1102